Michael Peter Schackert

Scanning Tunneling Spectroscopy on Electron-Boson Interactions in Superconductors

Experimental Condensed Matter Physics
Band 13

Herausgeber
Physikalisches Institut

Prof. Dr. Hilbert von Löhneysen
Prof. Dr. Alexey Ustinov
Prof. Dr. Georg Weiß
Prof. Dr. Wulf Wulfhekel

Eine Übersicht aller bisher in dieser Schriftenreihe erschienenen Bände finden Sie am Ende des Buchs.

Scanning Tunneling Spectroscopy on Electron-Boson Interactions in Superconductors

von
Michael Peter Schackert

KIT Scientific Publishing

Dissertation, Karlsruher Institut für Technologie (KIT)
Fakultät für Physik
Tag der mündlichen Prüfung: 6. Juni 2014
Referenten: Prof. Dr. W. Wulfhekel, Prof. Dr. J. Schmalian

Impressum

KIT Scientific Publishing

Karlsruher Institut für Technologie (KIT)
KIT Scientific Publishing
Straße am Forum 2
D-76131 Karlsruhe

KIT Scientific Publishing is a registered trademark of Karlsruhe Institute of Technology. Reprint using the book cover is not allowed.

www.ksp.kit.edu

Print on Demand 2014

ISSN 2191-9925
ISBN 978-3-7315-0238-8
DOI 10.5445/KSP/1000041865

Scanning Tunneling Spectroscopy on Electron-Boson Interactions in Superconductors

Zur Erlangung des akademischen Grades eines

DOKTORS DER NATURWISSENSCHAFTEN

von der Fakultät für Physik
des Karlsruher Instituts für Technologie

genehmigte

DISSERTATION

von

Dipl.-Phys. Michael Peter Schackert

aus Ulm

Datum der mündlichen Prüfung: 6. Juni 2014
Referent: Prof. Dr. W. Wulfhekel
Korreferent: Prof. Dr. J. Schmalian

Constantly regard the universe as one living being, having one substance and one soul; and observe how all things have reference to one perception, the perception of this one living being; and how all things act with one movement; and how all things are the cooperating causes of all things which exist; observe too the continuous spinning of the thread and the contexture of the web.

(Marcus Aurelius)

Contents

1 Introduction

Already today superconductors possess a wide range of applications [1]. Superconducting magnets produce the large magnetic fields of several tesla needed for magnetic resonance imaging (MRI) [2] or for particle accelerators [3]. Superconducting quantum interference devices (SQUID) are able to detect the minute magnetic fields produced by electrical currents occurring in the brain, thereby allowing mapping of the brain activity. Furthermore, superconducting quantum bits are promising candidates for building a quantum computer [4]. Because superconductivity only appears at temperatures well below room temperature, its technical use is currently limited to these high-tech applications where the enormous costs arising from the necessity of cooling the superconducting device by cryogenic liquids are justified by the lack of alternatives. Since the discovery of superconductivity in 1911, the quest for materials with higher transition temperatures has played an important role. A superconductor operating at ambient temperature, superseding the energy-demanding cooling process, would provide a huge, even more fascinating variety of applications, especially in terms of contributing to solutions to the growing task of power efficiency, starting from wires for loss-free electric power transmission and smaller (lighter) electric motors and generators. Additionally, large superconducting coils could provide very efficient power storage devices needed in order to mitigate the fluctuations in a power network comprising more and more non-continuous renewable energy sources. Among the imaginable applications are magnetic levitation vehicles which are trapped to a magnetic rail on which they move without friction [5].

Unfortunately, the search for new superconductors with higher transition temperatures made only slow progress that often appeared in an entirely unexpected direction. For example, the ground-breaking discovery of the so-called high temperature superconductors where the superconducting state could be achieved for the first time by much cheaper liquid nitrogen instead of liquid helium, happened in an insulating ceramic [6]. This finding was not foreseen by the theory of superconductivity available at that time. The main essence of this theory is the formation of electron pairs due to an effective attractive interaction between electrons which is mediated by bosonic exchange particles. In the case of conventional superconductors the exchange bosons were identified as phonons, the quantized vibrations of the crystal lattice. The name unconventional superconductor arose mainly from the theoretical finding that phonons are not able to cause the much higher transition temperatures [7]. Since, a variety of additional material classes possessing a pairing mechanism other than phonons have been found among which the iron-based pnictide superconductors, only discovered in 2008 [8], attract certainly most attention at present.

Despite tremendous efforts, to date there is still no generally accepted complete theoretical explanation for any of the unconventional systems. Though it is widely believed that still the formation of electron pairs is a prerequisite for superconductivity also in the novel materials, the question for the "glue" - the nature of the exchange boson - is not answered. Albeit magnetic fluctuations, present in all of these materials, are considered a promising candidate, an unambiguous experimental proof is not yet available. Anyhow, without a detailed knowledge of the microscopic origin, there is no way of tailoring materials with higher transition temperatures and thus the dream of superconductivity at room temperature further relies on unexpected discoveries by chance.

Unlike the situation of unconventional superconductivity, the details of the phonon as exchange particle in conventional superconductors are very well established. Eliashberg strong coupling theory provides an elaborate description of how the superconducting properties are influenced by phonons. For example, in the superconducting regime there is a distinct signature of phonons in the electronic density of states. The latter can be directly measured in a very elegant manner by electron tunneling spectroscopy. From such tunneling spectra a central quantity can be reconstructed [9], which is the effective phonon spectrum, also called Eliashberg function, $\alpha^2F(\omega)$, where $F(\omega)$ is the phonon density of states and α is the electron-phonon coupling strength. Apart from this rather indirect access to the Eliashberg function based on tunneling in the superconducting state, there exists a more direct way of probing phonons provided by another mode of tunneling spectroscopy: Based on the fact that the lattice vibrations are not an exclusive property of the superconducting state they can also be excited by electrons tunneling in the normal state. The signal produced by these inelastic processes then directly reflects the effective phonon spectrum.

Using planar tunnel junctions comprising an oxide barrier, tunneling spectroscopy on conventional superconductors was extensively employed in the 1960ies and early 1970ies, mainly in the superconducting state but also, though rarely, in the normal state. Nowadays the scanning tunneling microscope (STM) provides true vacuum tunneling. Further, it allows to perform tunneling spectroscopy with spatial resolution and hence the possibility of detecting local variations in the coupling of electrons to the exchange particles responsible for superconductivity. Surprisingly, experiments primarily focusing on the investigation of the electron-phonon coupling in conventional superconductors by means of the spectroscopic modes of the STM have not been reported in literature, neither in the superconducting nor in the normal state leaving the question open whether the early findings in planar tunnel junctions can be reproduced in the geometry of the STM. It thus appears necessary to make up these experiments, especially against the background of recent STM studies on unconventional superconductors which give rise to speculations on the existence of bosonic modes in the tunneling spectra of these materials.

This work presents the experimental results of scanning tunneling spectroscopy of the electron-boson interaction in conventional and unconventional superconductors. Chap. 2 gives an introduction to the main ideas of BCS theory and to its extension in the case of strong electron-phonon coupling before the deviating properties of unconventional superconductors are explained. The principles of scanning tunneling microscopy and tunneling spectroscopy - elastic and inelastic - are treated in Chap. 3 where, in the end, special attention is paid to the signatures of bosons in the tunneling spectra of superconductors. After describing the experimental setup (Chap. 4) the results of tunneling spectroscopy on conventional superconductors, namely on a niobium single crystal and on lead islands deposited on different substrates are shown in Chap. 5 and Chap. 6, respectively. The measurements on both systems demonstrate that the spectra in the superconducting state obtained by STM are in agreement with previous studies of planar tunnel junctions. After suppressing superconductivity by different means, signatures of phonon excitation are observed in inelastic tunneling spectra of each of the two systems. These features are very clear in the spectra of ultra-thin lead islands where, thanks to the spatial resolution of the STM, it was additionally found that the electron-phonon coupling interrelates with the energy of electronic quantum well states present in this system. Finally, Chap. 7 presents the preliminary results of experiments on a $SrFe_2(As_{1-x}P_x)_2$ crystal, a member of the 122 family of iron-based superconductors, as an outlook to possible future studies of the electron-boson coupling in unconventional superconductors.

The experimental findings on lead islands presented in Chap. 6 are supported by theoretical *ab initio* calculations performed by A. Ernst *et al.* at the Max-Planck Institut für Mikrostrukturphysik in Halle, Germany.

2 Electron-Boson Interaction in Superconductors

For many decades intensive research has been carried out in the field of superconductivity. In the meantime an immense number of review letters, monographs and textbooks (with different foci) were published. The first chapter of this thesis gives at first a short summary of the history and the fundamental properties of superconductivity. Advanced aspects are only treated as far as necessary to relate the concepts of this work. Slightly more attention is paid to the theoretical explanation of conventional superconductivity via electron-phonon interaction. The third part, finally, is dedicated to spin excitations as another type of bosonic quasiparticles and how they are thought to cause superconductivity in the high transition temperature materials.

2.1 Superconductivity: From the Discovery to the First Microscopic Theory

At room temperature the electrical resistance of a metal is mainly caused by electron-phonon and electron-electron scattering. One can thus easily understand its reduction upon decreasing temperature since the phonons freeze out and the phase space for electron-electron scattering becomes smaller. However, a residual resistance is always expected from the scattering at unavoidable lattice defects and impurities [7]. It was therefore very surprising when H. Kamerlingh Onnes discovered that, at a certain temperature, the resistance of mercury suddenly drops to an immeasurably small value when it is cooled in liquid helium [10]. Due to the perfect conductance the phenomenon was called superconductivity and the temperature below which it appears is referred to as critical temperature T_c. In 1913, only two years later, Kamerlingh Onnes was awarded the Nobel prize for his pioneering work in low-temperature physics. In the following, it was found that a lot of elements become superconducting, though many of them with a T_c below 1 K, and in some cases only upon applying a high pressure [1].

In 1933, W. Meißner and R. Ochsenfeld showed that superconductors are not only characterized by perfect conductivity but also by perfect diamagnetism [11]. The magnetic susceptibility being $\chi = -1$, superconducting materials expel magnetic fields completely from their interior ($B = 0$). Since this doesn't depend on the history of the sample, the superconducting state is a true thermodynamic one.

From the theoretical point of view it was at first only possible to set up phenomenological descriptions like the two-liquid model treating the superconductor as a mixture of normal conducting and superconducting charge carriers. In 1935, F. and H. London proposed a phenomenological model based on Maxwell's equations and capable to describe the electrodynamical properties of a superconductor [12]. They assumed a finite density n_S of "superconducting electrons" which starts to rise continuously from zero at T_c upon lowering the temperature.

In 1950, V. L. Ginsburg and L. D. Landau discussed the superconducting state in the framework of Landau's general theory of phase transitions. As the order parameter they introduced a macroscopic wave function Ψ and associated it with the local density of "superconducting electrons" ($n_S(\vec{r}) = |\Psi(\vec{r})|^2$) [7].The Ginsburg-Landau theory is especially suited for treating phenomena involving spatial variations of the superconducting charge carriers [7]. For example, A. Abrikosov showed on the basis of Ginsburg-Landau theory that one has to distinguish between type-I and type-II superconductors and, moreover, he could describe the vortex lattice in type-II superconductors [7]. While the London equations and the Ginsburg-Landau theory were and still are very helpful in order to describe many aspects of superconductivity, they are, however, due to the lack of a microscopic picture, not able to explain either the value of the critical temperature or the absence of dissipation in charge transport.

On the way towards a microscopic understanding of superconductivity it was realized that correlations in the electronic system play an important role in a superconducting solid. The theoretical treatment of correlated systems in general is challenging because the effective single particle picture, typically used in order to describe the dynamics of the crystal electrons, is not valid anymore [7]. For this reason it took almost 50 years after the discovery until, in 1957, J. Bardeen, L. N. Cooper, and J. R. Schrieffer succeeded in formulating a microscopic theory capable of describing the appearance of superconductivity [13]. For the BCS theory the inventors were awarded the Nobel prize in 1972. At the heart of BCS theory lies the finding that the Fermi sea, as the ground state of a normal metal, becomes instable as soon as there is any attractive interaction between the electrons. Considering the repulsive Coulomb interaction of the negatively charged electrons, the idea of a such an attractive interaction overcoming the electrostatic repulsion was certainly one of the most important steps.

Concerning the nature of the exchange particles mediating the attractive electron-electron interaction in superconductors, Bardeen, Cooper and Schrieffer had phonons[1] in mind when setting up their theory. This was motivated by the

[1] The theoretical description of phonons as quantized lattice vibrations will be introduced in the next section.

isotope effect [14] - the experimental observation that in many superconductors the transition temperature behaves as $T_c \propto 1/\sqrt{M}$, where M is the isotope mass and thus T_c depends on the frequency of the lattice vibrations. However, it is important to mention that the exact nature of the exchange particles certainly affects the details of the superconducting state, as will be demonstrated in the next section, but within the framework of BCS theory it does not act on the appearance of superconductivity in general. Therefore, any bosonic excitation in a solid capable of mediating an attractive interaction between the electrons might be considered, e. g. magnons, polarons or plasmons [7].

Cooper pairs and the BCS ground state

Within BCS theory the energy of two metal electrons subject to an attractive interaction can be calculated to be [7]

$$E_{\text{pair}} = 2E_F - \frac{2\,\hbar\omega_c\,\mathrm{e}^{-2/D(E_F)V_0}}{1 - \mathrm{e}^{-2/D(E_F)V_0}}, \tag{2.1}$$

where $D(E_F)$ is the electronic density of states per spin channel in the normal state at the Fermi level. V_0 is the strength of the interaction potential and $\hbar\omega_c$ a characteristic energy of the exchange bosons. The energy of the two interacting electrons is thus always smaller than $2E_F$, the energy of the two individual electrons, and therefore a bound pair is more favorable. At the basis of this derivation is a gedankenexperiment, where two electrons are artificially added to the edge of the Fermi sea at $T = 0$. In reality, all the electrons in a certain energy interval around E_F form so-called Cooper pairs which are bosons and are thus able to undergo a kind of Bose-Einstein condensation[2] [15] below a critical temperature and form a coherent many-body state - the BCS ground state [7].

Since, in BCS theory, the interaction potential is just given by a constant without any directional dependence the spatial part of the Cooper pair wave function has to be isotropic (total orbital momentum $L = 0$) as well. Due to Pauli's principle this requires an antisymmetric spin state ($S = 0$) and thus a Cooper pair is characterized by

$$\left\{\vec{k}\uparrow, -\vec{k}\downarrow\right\}, \tag{2.2}$$

a spin singlet with opposite momentum of the electrons. In all elemental superconductors this kind of Cooper pair is present. Different symmetry of the interaction potential in more complicated materials, however, may lead to a

[2]There are, however, serious differences between superconductivity and the superfluid phase transition, e. g. in ^{4}He. Cooper pairs only form below T_c and thus no "free" non-condensed bosons exist. Furthermore, Cooper pairs are not point-like particles but extend over a long distance and are therefore strongly correlated with other Cooper pairs.

pairing with $L \neq 0$ and thus eventually to a spin triplet as well [7, 16]. In phonon-mediated superconductors the size of a Cooper pair can be estimated to be of the order of 10 to $100\,\mathrm{nm}$. Since the electron density in a solid is about $10^{23}\,\mathrm{cm}^{-3}$, it is obvious that inside the volume of one Cooper pair there are a lot of other Cooper pairs (typically around 10^6), which means that they strongly overlap. This may help to imagine that a coherent quantum state will form [1]. According to BCS theory the latter is characterized by the coherence factors [7]

$$|v_{\vec{k}}|^2 = \frac{1}{2}\left(1 - \frac{\epsilon_{\vec{k}}}{\sqrt{\epsilon_{\vec{k}}^2 + |\Delta_{\vec{k}}|^2}}\right) \quad \text{and} \quad |u_{\vec{k}}|^2 = 1 - |v_{\vec{k}}|^2, \tag{2.3}$$

which give the probabilities of the state $|\vec{k}\uparrow; -\vec{k}\downarrow\rangle$ being occupied or unoccupied, respectively. The state $|\vec{k}\uparrow; -\vec{k}\downarrow\rangle$ is still describing single particles, but the two states potentially able to form a Cooper pair are treated together. The main conclusion of Eq. 2.3 is that, already at $T = 0$, the occupation probability of single-particle states is smeared around E_F (see Fig. 2.1(a)). Although this is at the cost of kinetic energy it allows the energy-gaining electron scattering processes, which would not be possible with a sharp border between occupied and unoccupied states [1]. Fig. 2.1(a) also shows the corresponding pair amplitude $g_{\vec{k}} = v_{\vec{k}} u_{\vec{k}}^*$, the density of Cooper pairs, only being non-zero around E_F. While, in Eq. 2.3, $\epsilon_{\vec{k}}$ is the single-electron energy with respect to the Fermi level of the normal metal, $\Delta_{\vec{k}}$ is identified with the superconducting order parameter. It is described by the BCS energy gap equation [7]

$$\Delta_{\vec{k}} = -\sum_{\vec{k}'} V_{\vec{k},\vec{k}'} \Delta_{\vec{k}'} \frac{\tanh(E_{\vec{k}'}/2k_BT)}{2E_{\vec{k}'}}. \tag{2.4}$$

As will be explained later, in BCS theory, the order parameter is just a real number ($\Delta_{\vec{k}} \to \Delta$) for which, at zero temperature, the quite simple expression

$$\Delta(T=0) \approx 2\,\hbar\omega_c\, \mathrm{e}^{-2/D(E_F)V_0} \tag{2.5}$$

can be derived, when $D(E_F)V_0 \ll 1$ is assumed, which is true for weak-coupling superconductors. Finally, the BCS theory yields a similar expression for the critical temperature

$$k_BT_c \approx 1.13\,\hbar\omega_c\, \mathrm{e}^{-2/D(E_F)V_0}, \tag{2.6}$$

and thus there exists a universal ratio of the gap value and the critical temperature

$$\frac{2\Delta(0)}{k_BT_c} \approx 3.53, \tag{2.7}$$

which can be used to test the theory experimentally.

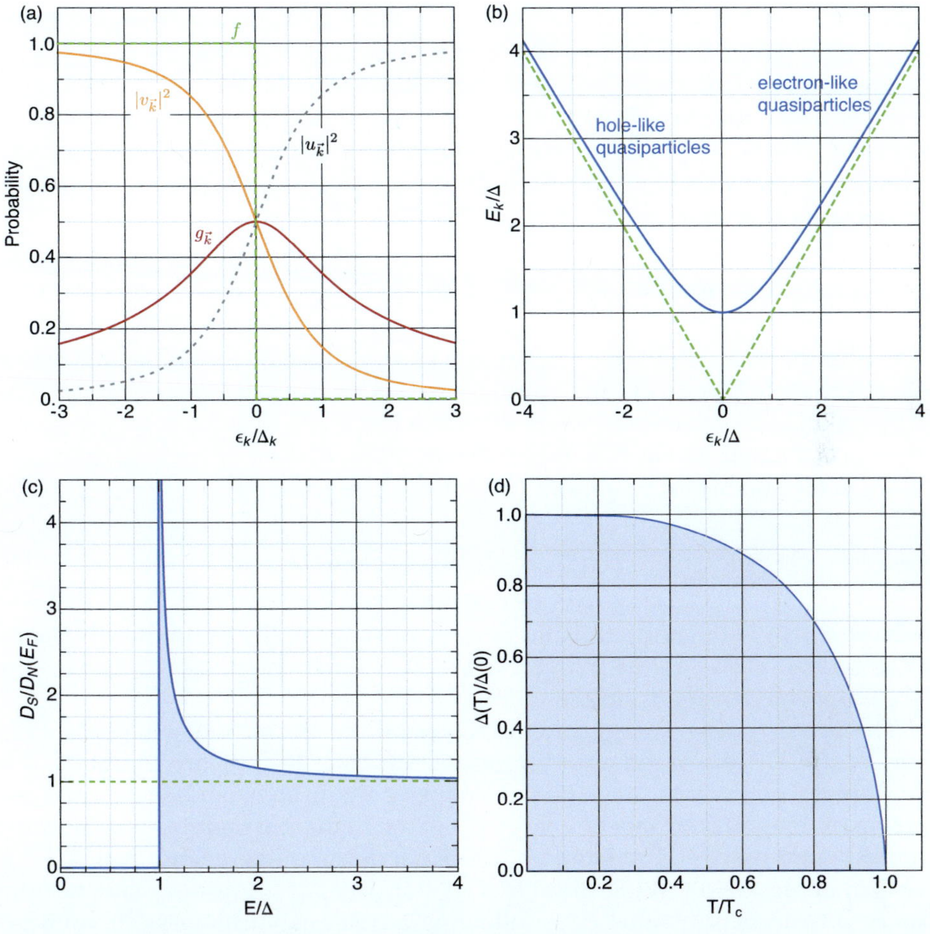

Figure 2.1: BCS ground state and quasiparticle excitations: (a) Coherence factors $|v_{\vec{k}}|^2$ (orange) and $|u_{\vec{k}}|^2$ (gray) together with the Fermi function f (green) and the pair amplitude $g_{\vec{k}}$ (red) in the BCS ground state at $T = 0$. $|v_{\vec{k}}|^2$ describes the occupation of pair states $|\vec{k}\uparrow; -\vec{k}\downarrow\rangle$ (after [7], page 846). (b) Comparison of the excitation energy of quasiparticles in a superconductor (blue) and electrons and holes in a normal metal (green). (c) Superconducting DOS (blue) and DOS of the normal state (green). (d) Temperature dependence of the superconducting gap (curve taken from [7], page 850).

Elementary excitations and the energy gap

The simplest excitation from the BCS ground state one can think of is a single unpaired electron $|\vec{k}, \uparrow\rangle$. This excitation is more complicated than it appears at the first glance because one has to assure conceptually that the state $|-\vec{k}, \downarrow\rangle$ is unoccupied. Otherwise, a Cooper pair would form immediately. The single-particle excitations in a superconductor are called quasiparticles, as are elementary excitations from a many-body ground state in general [17]. The quasiparticles are characterized by $E(\vec{k})$, their dispersion relation, which is derived from BCS theory to read as

$$E(\vec{k}) = \sqrt{\epsilon_{\vec{k}}^2 + \Delta^2} \quad \text{with} \quad \epsilon_{\vec{k}} = \frac{\hbar^2 k^2}{2m} - E_F. \tag{2.8}$$

The quasiparticles thus have a minimum energy given by the value of Δ which therefore describes an energy gap for excitations from the ground state. The dispersion in the vicinity of E_F is shown in Fig. 2.1(b) together with the one of the normal metal. In Fig. 2.1(c) the corresponding density of states of quasiparticles in the superconducting state is plotted. It is given by

$$D_S(E) = D_N(0)\frac{|E|}{\sqrt{E^2 - \Delta^2}}, \tag{2.9}$$

where $D_N(0)$ is the normal metal density of states at the Fermi level. $D_S(E)$ is zero for $E < |\Delta|$. While it diverges at $\pm\Delta$, the decaying edges approach the normal metal density of states.

The opening of an energy gap in the excitation spectrum of a superconductor is certainly one of the most important results of BCS theory. Actually, it is this energy gap which prevents the Cooper pairs from experiencing any scattering (as long as their kinetic energy is not too high) and therefore explains the perfect conductivity. The temperature dependence of the energy gap $\Delta(T)$ is shown in Fig. 2.1(d). While close to $T = 0$ it decreases only very slowly from its zero temperature value $\Delta(0)$ following a non-analytical curve, it vanishes like $\sqrt{1 - T/T_c}$ at T_c [7]. The energy gap was observed experimentally in absorption spectroscopy using infrared light or ultrasound as well as in specific heat measurements, but the most direct and elegant method is quasiparticle tunneling, which will be discussed in detail in Sec. 3.3.

2.2 Phonon-Mediated Superconductivity

As explained in the previous section the basic idea of BCS theory is the formation of a new ground state made up by Cooper pairs which are formed due to an attractive interaction mediated by the exchange of virtual bosons. Although

BCS theory was originally formulated for phonons, for the sake of generality, the discussion so far was intentionally kept free of a distinct sort of exchange boson. This section will now focus on the details of phonon-mediated superconductivity and show under which assumptions the electron-phonon coupling enters BCS theory. Though the predictions of the latter were confirmed in numerous experiments on metals and alloys, some deviations were found in the case of so-called strong-coupling superconductors, which are characterized by a violation of Eq. 2.7, i. e. $2\Delta/k_BT_c > 3.53$, the value obtained by BCS theory. The deviations were found to arise from the detailed nature of the electron-phonon coupling which is included in Eliashberg theory. As a starting point, however, the theoretical description of phonons will be summarized.

2.2.1 Theoretical Description of Phonons

In every solid the atoms can be excited to perform oscillations about their equilibrium positions. The latter is given by the minimum of the electrostatic potential between the atoms. In an ordered crystal the translation symmetry allows the quantitative description of lattice vibrations. Although these lattice vibrations can be understood quite well as collective excitation modes already from a classical point of view, further insight is possible using a quantum mechanical treatment in the second quantization. In this picture the lattice vibrations are represented by bosonic quasiparticles called phonons.

The starting point for the theoretical treatment of vibrating atoms in a solid is the Hamilton operator

$$H = \sum_{l=1}^{N} \frac{\vec{P}_l^2}{2M_l} + V_{\text{eff}}(\vec{R}_1^{(0)} + \vec{u}_1, \; ..., \; \vec{R}_N^{(0)} + \vec{u}_N) \tag{2.10}$$

for N atoms of mass M_l at positions $\vec{u}_l$ with respect to their equilibrium positions $\vec{R}_l^{(0)}$, all interacting with each other via the effective potential V_{eff}. The procedure how to simplify the potential using the harmonic approximation in order to arrive at the dynamical matrix comprising the eigenmodes ω_l of $3N$ independent oscillators can be found in textbooks about solid state physics. A very elaborate treatment is presented in [15]. There it is also explained in detail, that, in an ordered crystal, one can make use of the translational symmetry by shifting the problem into momentum space which reduces the computational demand drastically (or rather allows a calculation at all). Finally, after introducing $b^\dagger_{j\vec{q}}$ and $b_{j\vec{q}}$, the bosonic creation and annihilation operators, respectively, one ends up with the Hamilton operator

$$H_{\text{harm}} = \sum_{\vec{q}} \sum_{j}^{3r} \hbar\omega_j(\vec{q}) \left(b^\dagger_{j\vec{q}} b_{j\vec{q}} + \frac{1}{2} \right), \tag{2.11}$$

describing a system of independent oscillators, whose energies are given by the dispersion relations $\omega_j(\vec{q})$. Concerning the polarizations j, in a three-dimensional crystal comprising only one atomic species there are one longitudinal and two transversal phonon modes. For small momentum, the dispersion of these modes is linear in q, which are therefore called acoustic phonons. In more complicated crystals there are additional modes possessing a non-vanishing energy at $q = 0$ referred to as optical phonons. In a lattice with r atoms in its basis there are 3 acoustic and $3(r-1)$ optical branches of the dispersion relation.

Very often, for practical purposes, the full information provided by the dispersion relations $\omega_j(\vec{q})$ is not necessary. Many physical properties indeed only depend on the energy of the lattice vibrations and not explicitly on their momentum and thus it is sufficient to know how many phonons exist at a certain energy. This information is given by the phonon density of states $F(\omega)$ which is derived from the dispersion relation as [15]

$$F(\omega) = \frac{V}{N} \sum_j \frac{1}{(2\pi)^3} \int_{S(\omega)} \frac{\mathrm{d}s}{|\nabla_{\vec{q}}\,\omega_j(\vec{q})|}, \tag{2.12}$$

where the prefactors take care of correct normalization. The important step is the integration in reciprocal space over a surface of distinct energy in order to arrive at the average of all phonons with the corresponding energy, and the directional information is lost.

As an example, the dispersion relation and the resulting density of states of a three-dimensional simple cubic system with one atom in the primitive cell is illustrated in Fig. 2.2. The dispersion relation is plotted along some selected directions. Since there is only one atom in the Wigner-Seitz cell no optical phonons appear but only the three acoustic modes. Furthermore, along high symmetry directions, the transversal modes are degenerate. This simple example is also suited for giving an impression that the shape of dispersion relations is dominated by the given crystal structure. At some points the phonon energy does not change with momentum and the dispersion relation is flat implying $\nabla_{\vec{q}}\,\omega_j(\vec{q}) = 0$. At the corresponding energies the density of states diverges resulting in characteristic peaks which are referred to as van Hove singularities. A further attribute of all phonon densities of states is the low energy part which is quadratic in energy due to the linear dispersion of acoustic phonons for small values of q.

The most powerful experimental technique for determining the phonon dispersion relation of a crystal is certainly inelastic neutron scattering (INS) because the energy and the momentum of neutrons are just of the right order of magnitude [18]. Neutrons impacting on a crystal are therefore able to absorb or emit phonons during the scattering process and the corresponding energy and momentum transfer can be measured from the scattered neutrons. Results of INS on materials used in this study are presented in the respective chapters.

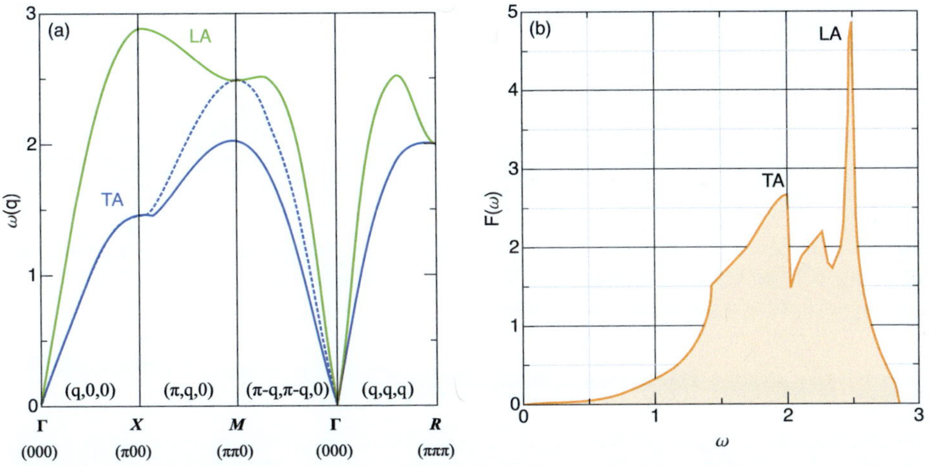

Figure 2.2: Phonons in a simple cubic lattice: (a) Dispersion relation along certain directions. Longitudinal acoustic (LA, green) and transversal acoustic (TA, blue) branches. The latter two are degenerate along high symmetry directions. (b) Resulting DOS exhibiting characteristic van Hove singularities mainly arising from the indicated branches. (curves from [15, page 63 and 67])

In order to allow for the deduction of thermodynamic properties of a solid using simple calculations, Debye introduced a model, valid for crystals with a one-atomic basis, which drastically simplifies the dispersion relation. The Debye model assumes an isotropic linear dispersion $\omega(\vec{q}) = c_s q$, where c_s is an appropriate sonic velocity. The linear dispersion is cut off at a certain wave number $q_D = \sqrt[3]{6\pi^2 N/V}$ simply determined by the density of atoms in the crystal. One usually defines a Debye temperature T_D and a Debye frequency ω_D in the following form:

$$k_B T_D = \hbar\omega_D = \hbar\, c_s q_D. \tag{2.13}$$

The Debye temperature is determined by matching specific heat measurements. Typical values are 100 to 1000 K corresponding to a Debye frequency $\hbar\omega_D$ of about 10 meV to 100 meV [15].

Phonons are very important for many processes in a solid, e. g. thermal conductivity or electrical resistance via electron-phonon scattering. Since the electron-phonon interaction is also essential for the understanding of conventional superconductivity it will be discussed in the following.

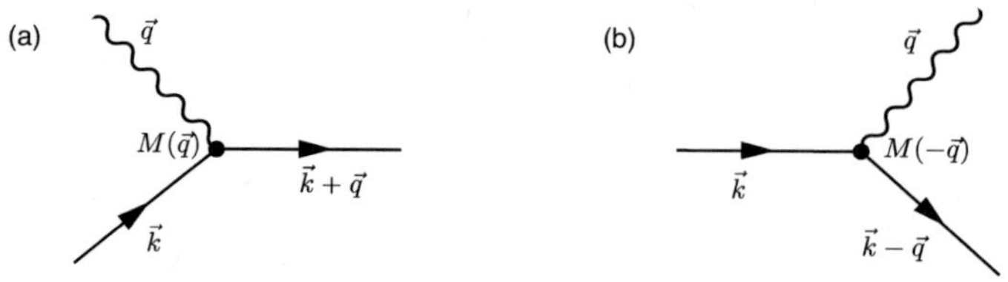

Figure 2.3: Feynman diagrams of electron-phonon scattering: (a) Absorption of a phonon. (b) Emission of a phonon. The momentum at the vertex is actually only conserved except for a reciprocal lattice vector.

2.2.2 Electron-Phonon Interaction

So far, the phonons were treated as a free system without any interaction with the electrons justified by the very different masses of electrons (m) and atoms (M) allowing to describe the dynamics of the lattice and of the electronic system independently in the Born-Oppenheimer approximation [7]. However, corrections to this approximation which are classified by powers of $\sqrt{m/M}$ may become important. The electron-phonon interaction is conveniently treated in second quantization using fermion and boson creation and annihilation operators, $c^{(\dagger)}_{\vec{k},\sigma}$ and $b^{(\dagger)}_{\vec{q},j}$, for electrons and phonons, respectively. Since the phonon scattering does not influence the spin of an electron, σ is omitted. The Fröhlich Hamilton operator [15] for the electron-phonon interaction then reads

$$
\begin{aligned}
H_F = & \sum_{\vec{k}} \varepsilon(\vec{k})\, c^{\dagger}_{\vec{k}} c_{\vec{k}} \\
& + \sum_{\vec{k},\vec{G}} \sum_{\vec{q},j} \left[M_j(\vec{q}) c^{\dagger}_{\vec{k}+\vec{q}+\vec{G}} c_{\vec{k}} b_{\vec{q},j} + M_j(-\vec{q}) c^{\dagger}_{\vec{k}-\vec{q}+\vec{G}} c_{\vec{k}} b^{\dagger}_{\vec{q},j} \right] \qquad (2.14) \\
& + \sum_{\vec{q},j} \hbar\omega_j(\vec{q}) \left(b^{\dagger}_{\vec{q},j} b_{\vec{q},j} + \frac{1}{2} \right) .
\end{aligned}
$$

While the first and the third term in Eq. 2.14 refer to free electrons and free phonons, respectively, the second one, where $M_j(\vec{q})$ is the matrix element of the electron-phonon interaction, describes the scattering of electrons from Bloch states with wave vector $\vec{k}$ into states $\vec{k}+\vec{q}$ absorbing (annihilating) a phonon of momentum $\hbar\vec{q}$ or emitting (creating) one leaving the electron in the state $\vec{k}-\vec{q}$ as illustrated by the Feynman diagrams in Fig. 2.3. The momentum at the vertex is conserved except for a reciprocal lattice vector $\vec{G}$. The phonon emitted by one electron can be absorbed by another electron and therefore transmits an interaction. That this interaction can be attractive may be understood vividly by considering the polarization of positively charged lattice ions induced by a passing electron. Since the electrons move much faster than the heavy ions, a second electron will feel an attraction at a later time.

Using a suitable canonical transformation one can bring the interaction term in Eq. 2.14 to a form in which the phonon operators do not appear explicitly anymore in order to end up with [15]

$$H_{\text{e-e}} = \sum_{\vec{k},\vec{k}'} \sum_{\vec{q}} |M(\vec{q})|^2 \frac{\hbar\omega_{\vec{q}}}{(\varepsilon_{\vec{k}+\vec{q}} - \varepsilon_{\vec{k}})^2 - \hbar^2\omega_{\vec{q}}^2} c^{\dagger}_{\vec{k}'+\vec{q}} c_{\vec{k}'} c^{\dagger}_{\vec{k}-\vec{q}} c_{\vec{k}}. \tag{2.15}$$

For the sake of clarity in this expression only one branch of the dispersion relation is considered and umklapp processes ($\vec{G} \neq 0$) are neglected. Equation 2.15 can now be regarded as an effective interaction between two electrons by the exchange of a phonon of momentum $\hbar\vec{q}$ as illustrated in Fig. 2.4(a). The new matrix element becomes negative if $|\varepsilon_{\vec{k}+\vec{q}} - \varepsilon_{\vec{k}}| < \hbar\omega_{\vec{q}}$. Thus the electron-electron interaction mediated via virtual phonons becomes attractive in a narrow shell around the Fermi surface.

Electron-phonon interaction not only has drastic consequences like electrical resistivity or superconductivity but also leads to a more subtle effect, which is the change of the electronic band structure (and with this the density of states) close to the Fermi level. An advanced many-body treatment, the details of which are beyond the scope of this experimental thesis, shows that the "bare" electron energy $\varepsilon(\vec{k})$ has to be replaced by the full energy

$$E(\vec{k}) = \varepsilon(\vec{k}) + \Sigma(\vec{k}, E), \tag{2.16}$$

where the self-energy $\Sigma(\vec{k}, E) = \Sigma_{\text{Re}}(\vec{k}, E) + \Sigma_{\text{Im}}(\vec{k}, E)$ accounts for the correction to the band energy due to the interactions with phonons [19]. One of the consequences is for example that the "dressed" electrons have a higher band mass and finite lifetime broadening.

2.2.3 From Weak to Strong Coupling

Based on Eq. 2.15, the BCS interaction potential, which was introduced already in Eq. 2.1, can now be defined properly. It arises from the substitution

$$|M(\vec{q})|^2 \frac{\hbar\omega_{\vec{q}}}{(\varepsilon_{\vec{k}+\vec{q}} - \varepsilon_{\vec{k}})^2 - \hbar^2\omega_{\vec{q}}^2} \rightarrow \begin{cases} -V_0, & \text{if } 0 < |\varepsilon_{\vec{k}+\vec{q}} - \varepsilon_{\vec{k}}| < \hbar\omega_D, \\ 0, & \text{else.} \end{cases} \tag{2.17}$$

and drastically simplifies the $\vec{q}$-dependent electron-phonon coupling by assuming a constant isotropic interaction potential V_0. The original interaction, which is local in space and retarded in time is thus replaced by a non-local instantaneous one. The fact that it is mediated by phonons enters the whole theory only

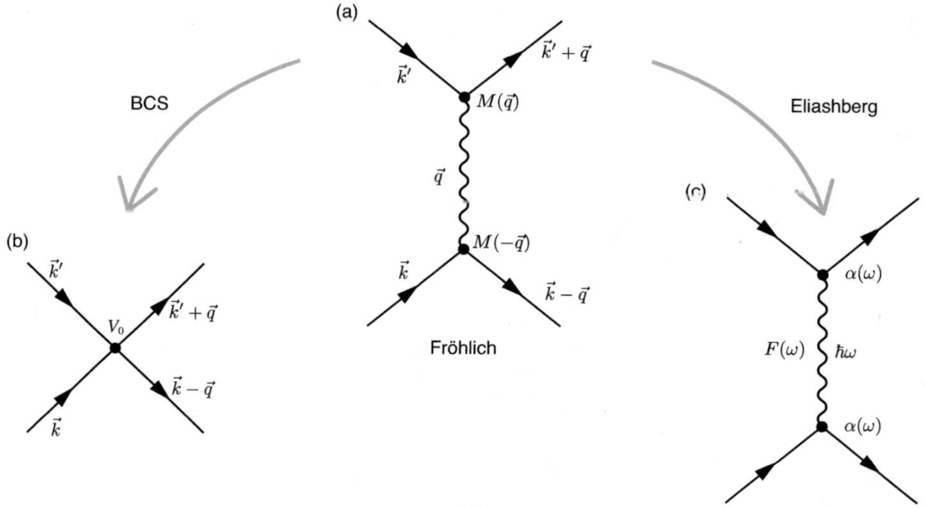

Figure 2.4: Illustration of electron-electron interaction in different models: (a) In the accurate Fröhlich model two electrons of momentum $\hbar\vec{k}$ and $\hbar\vec{k}'$ interact with each other by exchanging a phonon of momentum $\hbar\vec{q}$. (b) BCS theory summarizes all the scattering events in one effective interaction potential V_0. (c) Eliashberg theory also simplifies the actual Fröhlich interaction, but not as severely, by averaging over all phonons of the same energy.

since the Debye frequency ω_D is used as the cut-off for the potential. The corresponding Feynman diagram is shown in Fig. 2.4(b). Despite the rather crude simplifications the BCS theory was confirmed in many experiments, at least in the weak-coupling regime ($D(E_F)V_0 \ll 1$).

Once the main idea of electron pairing due to the attractive phonon mediated interaction was established within BCS theory, it was Eliashberg in 1960 [20] who presented a theory staying closer to the Fröhlich interaction potential (Eq. 2.15). The central quantity in this theory is the effective electron-phonon coupling $\alpha^2F(\omega)$, also called Eliashberg function, which is a dimensionless quantity defined as [19]

$$\alpha^2F(\omega) = \int_{S_F} \frac{\mathrm{d}^2k}{v_F} \int_{S'_F} \frac{\mathrm{d}^2k'}{(2\pi)^3 v'_F} \sum_j M_j(\vec{k}-\vec{k}\,')\delta(\omega-\omega_j(\vec{k}-\vec{k}\,')) \left(\int_{S_F} \frac{\mathrm{d}^2k}{v_F}\right)^{-1} . \tag{2.18}$$

It consists of the phonon density of states $F(\omega)$ and $\alpha = \alpha(\omega)$,[3] the mean value of the electron-phonon matrix element averaged over all phonons of energy $\hbar\omega$.

[3]The notation $\alpha^2F(\omega)$ omitting the ω dependence of α is just an abbreviation commonly found in literature.

The idea of the effective electron-electron interaction of the Eliashberg theory is illustrated in Fig. 2.4(c). The Eliashberg function is also an averaged quantity not distinguishing between different polarizations and not depending explicitly on the phonon momentum $\hbar\vec{q}$. However, still including the retardation and the spatial locality of the interaction as well as the true phonon density of states, it is a less strong simplification compared to the assumptions of BCS theory and thus represents essentially an extension of the latter.

The superconducting properties are obtained from the Eliashberg gap equations, which are given here in their zero temperature limit and with $\hbar = 1$ [19]:

$$\Delta(\omega) = \frac{1}{Z_S(\omega)} \int_0^{\omega_c} \mathrm{d}\nu \mathrm{Re}\left\{ \frac{\Delta(\nu)}{[\nu^2 - \Delta^2(\nu)]} \right\} [K_+(\nu,\omega) - \mu^*] \tag{2.19}$$

$$[1 - Z_S(\omega)]\,\omega = \int_0^{\infty} \mathrm{d}\nu \mathrm{Re}\left\{ \frac{\nu}{[\nu^2 - \Delta^2(\nu)]} \right\} K_-(\nu,\omega), \tag{2.20}$$

This is a system of complex integral equations for the gap parameter $\Delta(\omega)$, which now specifically depends on energy. $Z_S(\omega)$ is a normalization factor related to the self-energy in the superconducting state. μ^* is an electron-electron pseudo-potential, a parameter accounting for the screened but still repulsive Coulomb interaction. The integration runs to ω_c, a cut off frequency of the order $10\omega_D$. The Eliashberg function enters via

$$K_{\pm}(\nu,\omega) = \int_0^{\infty} \mathrm{d}\omega' \alpha^2 F(\omega') \left(\frac{1}{\omega + \omega' + \nu + \mathrm{i}\delta} \mp \frac{1}{\omega - \omega' - \nu + \mathrm{i}\delta} \right). \tag{2.21}$$

Thus, from the knowledge of the Eliashberg function (and the pseudo-potential) it is possible to derive the properties of conventional phonon-mediated superconductors, not only in the weak-coupling limit but also in the presence of strong electron-phonon interaction.

In order to get a feeling for the implications of the rather complicated theory it is instructive to have a look on its application to a simple example. Such a calculation was presented by Scalapino *et al.* [21] who solved the Eliashberg gap equations for a single phonon peak at ω_0 as depicted in Fig. 2.5(a). The resulting complex gap parameter $\Delta(\omega)$ is shown in Fig. 2.5. At low energies the real part $\Delta_{\mathrm{Re}}(\omega)$ has a constant value of Δ_0 which is actually the energy gap one would observe in experiment. Before $\Delta_{\mathrm{Re}}(\omega)$ changes from Δ_0 to zero at high energies, including an overshoot to negative values, it peaks at $\omega_0 + \Delta_0$, the energy of the phonon shifted by the gap value. The imaginary part $\Delta_{\mathrm{Im}}(\omega)$ exhibits only a peak slightly above $\omega_0 + \Delta_0$. Concerning the quasiparticle density of states in the strong coupling regime, Schrieffer *et al.* proposed that it should be given by [21, 22]

$$\frac{D_S(E)}{D_N(E_F)} = \mathrm{Re}\left(\frac{E}{\sqrt{E^2 - \Delta^2(E)}} \right), \tag{2.22}$$

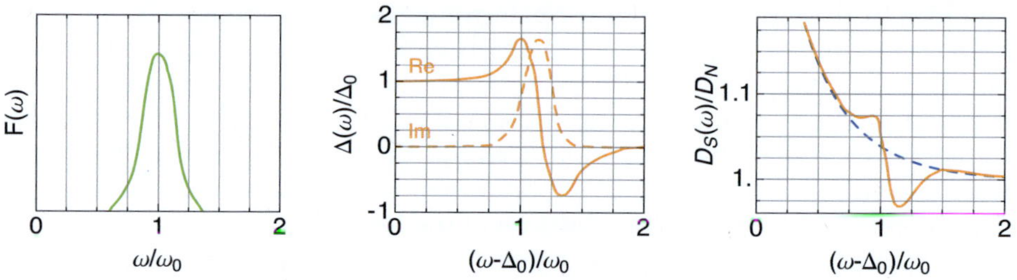

Figure 2.5: Application of the Eliashberg gap equations to a single-phonon peak model: (a) Assumed phonon density of states $F(\omega)$. (b) Real part (solid line) and imaginary part (dashed) of the energy-dependent gap parameter $\Delta(\omega)$. (c) The quasiparticle density of states, which is smoothly decaying in the BCS theory (blue) is altered around the energy of the phonon shifted by the gap value Δ_0. Curves taken from [21].

similar to the BCS expression besides the fact that Δ is not any more a constant but a complex, energy-dependent function. The result for the single-phonon peak model is presented in Fig. 2.5(c). Unlike the smoothly decaying BCS form, there are deviations from the latter around the energy of the phonon, again shifted by the gap. Especially, there is a sharp decrease very close to $\omega_0 + \Delta_0$.

Beyond this illustrative example, Eliashberg strong coupling theory is able to precisely explain similar deviations from the BCS quasiparticle density of states experimentally observed before in tunneling spectra of real materials [23], which will be presented in Sec. 3.4. The possibility of theoretically describing even minor details like the fine structure in the quasiparticle density of states lead to the general belief that the phenomenon of superconductivity is essentially understood, although a true "first-principle" calculation for the superconducting properties remained difficult even with Eliashberg theory. A problem is especially the determination of $\alpha^2 F(\omega)$ because the electron-phonon coupling and thus α is not easily accessible [15]. Nevertheless, Eliashberg theory allowed to estimate an upper limit for the critical temperature of phonon-mediated superconductors, which should be restricted to values below 40 K [7, 24].

2.3 Unconventional Superconductivity

Together with the extension by the Eliashberg strong coupling theory, leaving the basic ideas of BCS unaffected, it was widely believed that the phenomenon of superconductivity was understood in general. This changed suddenly when J. G. Bednorz und K. A. Müller discovered superconductivity in ceramics comprising copper oxide layers in 1986 [6] wherefore they were awarded the Nobel prize in 1987. Since the maximal T_c of these cuprates could be rapidly increased to more than 130 K [25], they are called high-temperature superconductors. As mentioned in the last section, these high transition temperatures cannot be explained in terms of conventional BCS theory, even with the extension of Eliashberg. While there is a general agreement that unconventional superconductivity is also achieved through the quantum condensing of Cooper pairs, to this day, however, the mechanism leading to their formation in the cuprates is still not conclusively understood. The same is true for superconductivity in heavy fermion systems discovered by F. Steglich in 1979 [26] as well as for iron pnictides[4]. Superconductivity at surprisingly high temperatures up to 56 K [27, 28] in this material class was only found in 2008 in the group of Hosono [8]. Because of the topicality of the subject also in this work an iron-based superconductor was chosen for the first experiments on unconventional superconductors and therefore much of the following discussion is focused on this type of material.

Superconductivity in a material containing magnetic moments was very surprising, since, according to the conventional theory, the presence of magnetic moments leads to scattering processes destroying the spin-singlet Cooper pairs. Indeed, alloying a vanishing small amount of magnetic atoms into a conventional superconductor causes superconductivity to disappear [1]. Attempts of tracing back the electron paring to a phonon-mediated interaction failed. For example, the application of the Eliashberg formalism to $LaFeAsO_{1-x}F_x$ as one example of iron pnictides leads to a relatively small electron-phonon coupling constant $\lambda = 0.21$ resulting in a transition temperature of $T_c = 0.8\,K$ deviating drastically from the measured $T_c = 26\,K$ in this material [29, 31]. As an alternative, due to the vicinity of a magnetically ordered phase (antiferromagnetic or spin density wave) to the superconducting region in the stoichiometric phase diagram of iron pnictides and also of cuprates, it has been speculated that spin fluctuations could take the role of mediating the attractive interaction between the electrons to form Cooper pairs [32]. Fig. 2.6 exemplarily shows the phase diagram of the electron doped compound $Ba(Fe_{1-x}Co_x)_2As_2$. As for most of these structures the undoped parent compound exhibits an antiferromagnetic order. By doping electrons, the latter will be suppressed and a superconducting dome will emerge [29]. Though conceptually very different compared to phonons as vibrational excitations of the crystal atoms, spin fluctuations can also be regarded as bosonic quasiparticles. Despite careful formulations in the relevant literature [33, 34] it seems that it is widely believed that spin fluctuations provide the "glue" for Cooper pair formation. Indeed, there are also experiments supporting this scenario. For example, the Eliashberg function was extracted from optical measurements on $Ba(Fe_{0.92}Co_{0.08})_2As_2$ and remarkable agreement with inelastic magnetic neutron scattering measurements was

[4]Compounds of the nitrogen group are collectively called pnictides

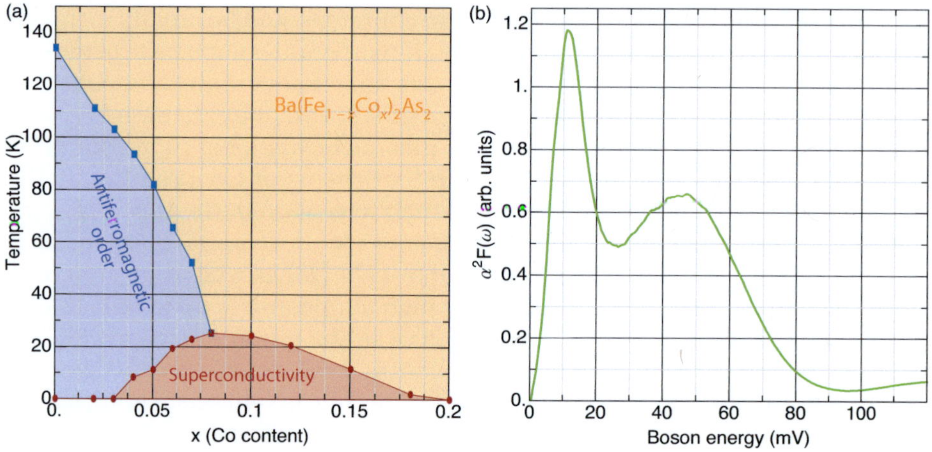

Figure 2.6: Evidence for spin fluctuations as pairing glue in iron pnictides at the example of $Ba(Fe_{1-x}Co_x)_2As_2$: (a) Stoichiometric phase diagram. Upon electron doping through substituting iron by cobalt the antiferromagnetic order gets suppressed and superconductivity emerges. From [29]. (b) Eliashberg function extracted from optical measurements. From [30].

found, indicating that spin fluctuations strongly couple to the electrons [30].

In contrast to conventional superconductors where the pair state is often referred to as "s-wave" because of the spin of a Cooper pair having $S = 0$ and its angular momentum $L = 0$, the different pairing mechanism in unconventional superconductors also potentially allows different symmetries of the order parameter [16]. Staying at a spin singlet for example would also permit $L = 2$, a "d-wave" symmetry, which is indeed realized in the cuprates [7]. As a consequence, the gap parameter is not isotropic, but now depends on the crystal directions for some of which it actually vanishes [35]. In those directions no minimum energy for quasiparticle excitation is required. Concerning the quasiparticle density of states, which reflects an average over the whole k space, this results in a V-shape around the Fermi level instead of the fully-gapped form in Fig. 2.1(c).

In the case of Fe-based superconductors yet another pairing symmetry is proposed: the $s^{\pm}$-symmetry. The gap parameter is isotropic but has different sign in the middle and at the edge of the Brillouin zone [32]. Depending on whether the surface on which Δ changes its sign intercepts the Fermi surface or not, this special symmetry will result in a V-shaped or a fully-gapped quasiparticle density of states, respectively.

Furthermore if there are different sheets of the Fermi surface, there may be different values of the order parameter on each of them. This so-called multiband superconductivity results in even more complicated structures in the quasiparticle density of states.

3 Scanning Tunneling Microscopy and Tunneling Spectroscopy

The experiments of this study were performed using a scanning tunneling microscope. However, electron tunneling provides a spectroscopic tool that was already widely exploited especially in the field of superconductivity long before the application in microscopy was suggested. This chapter is dedicated to the historical development and the theoretical description of electron tunneling experiments in order to relate the concepts of scanning tunneling microscopy and electron tunneling spectroscopy and, of course, the combination of both. In the end, there will be an introduction to inelastic tunneling spectroscopy and how it allows to detect the Eliashberg function in a very direct way - the main purpose of the present work.

3.1 Quantum Mechanical Tunneling of Electrons

3.1.1 History

Still in the 1920ies, very shortly after establishing quantum mechanics, physicists were able to apply its novel concepts to explain effects like the α-decay of nuclei, the field ionization of hydrogen and field emission from a metal surface by a phenomenon which was named "tunneling". This term relates to the fact that a small-mass particles can reach the other side of a potential barrier even though it has not enough energy in order to classically overcome it.

The origin of the tunneling phenomena is thus of purely quantum mechanical nature. The probability density of finding a particle of kinetic energy E at a certain position $\vec{r}$ is given by the square of the absolute value of its wave function $\Psi(\vec{r})$, which, in the stationary state, can be obtained by solving the time-independent Schrödinger equation

$$-\frac{\hbar^2}{2m}\frac{\partial^2\Psi(\vec{r})}{\partial\vec{r}^2} + V(\vec{r})\Psi(\vec{r}) = E\Psi(\vec{r}) \tag{3.1}$$

for a given potential $V(\vec{r})$. A very simple one-dimensional case of two regions, where $V(x) = 0$, separated by a square-shaped potential barrier of height V_0 and width d is shown in Fig. 3.1. Solving Eq. 3.1 in the regions left and right of the barrier, where $E > V_0$, one obtains standing waves with wave number $k = \sqrt{2mE}/\hbar$. In the classically forbidden region inside the barrier the solution

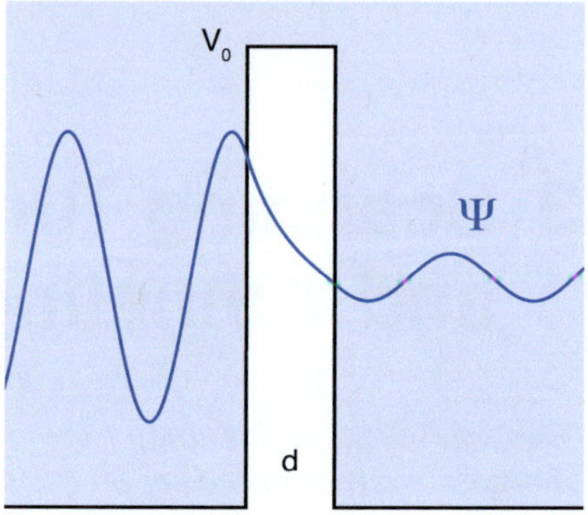

Figure 3.1: Tunneling through a square potential well: The amplitude of the wave function (only the real part of Ψ is plotted) of an electron coming in from the left is damped in the barrier region but still there is a finite possibility of finding the particle on the right.

is an evanescent wave with decay length $\kappa = \sqrt{2m(V_0 - E)}/\hbar$. Matching the solutions in such a way that $\Psi(x)$ and $\mathrm{d}\Psi/\mathrm{d}x$ are continuous at the edges of the barrier yields the amplitude t of a wave which initially comes in from the left after transmission through the barrier. For $\kappa d \gg 1$ one finds

$$T = |t|^2 = \frac{16k^2\kappa^2}{(k^2+\kappa^2)^2} \exp(-2\kappa d). \tag{3.2}$$

From the exponential dependence it is obvious that the probability for a particle to "tunnel" through the barrier strongly depends on the width of the latter [36, 37].

Having found this, it is not surprising that, even though the possibility that an electron current could flow between to metals separated by an insulating layer was considered already in 1930 [36], it took some more decades until the first working metal-insulator-metal (MIM) tunnel junctions were presented by J. C. Fisher and I. Giaever in 1959 [38]. Yet before, it was L. Esaki who managed to study electron tunneling in a solid-state structure for the first time by investigating narrow semiconductor *p*-*n* junctions. Fisher and Giaever, instead, kept focusing on MIM junctions and, finally, they were able to fabricate reproducible

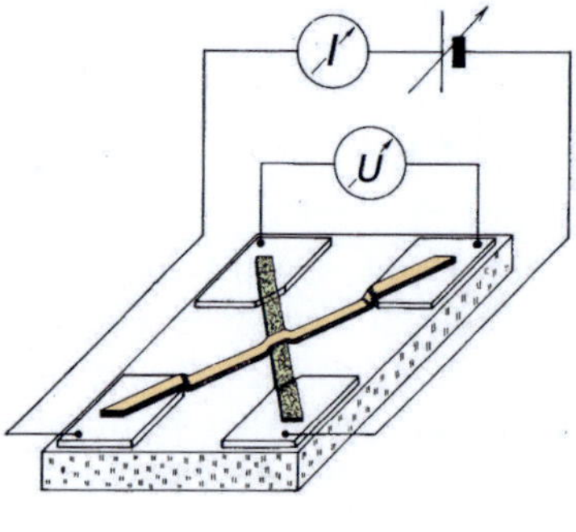

Figure 3.2: Planar tunnel junction: Long before the invention of the STM, tunneling measurements were already performed using planar tunnel junctions. The lower electrode is thermally oxidized forming a thin insulating oxide layer on the freshly evaporated metal. Subsequently, a crossing counter electrode is fabricated [1].

barriers of controlled thickness by exposing a freshly evaporated metal film to atmosphere in order to form a native oxide layer and, afterwards, depositing a crossing electrode on top [39] as depicted in Fig. 3.2.

Shortly after having demonstrated that a tunneling barrier can be successfully obtained as was just described, Giaever discovered that, when one electrode of such a junction is made of Al or Pb, the I-U characteristic shows a strong nonlinearity below the superconducting transition temperatures of these materials [40]. In particular it was found that there is no current flowing at all below an onset voltage $U = \Delta/e$, where Δ is the energy gap of the superconducting electrode.[1] Moreover it was discovered [40, 41] that the tunneling current contains detailed information about the quasiparticle density of states. "This discovery was to initiate a long and fruitful connection between tunneling and superconductivity, and to demonstrate the simplicity and elegance of tunneling as a spectroscopic tool." [36]. To date, tunneling experiments on the superconducting quasiparticle density of states are considered as one of the major confirmations of BCS and Eliashberg theory. In 1973, I. Giaever together with L. Esaki and B. D. Josephson[2] were awarded the Nobel prize for their work on tunneling phenomena in solid-state junctions.

3.1.2 Bardeen Model

The most extensively used theoretical model for the understanding of the tunneling process in solid-state structures is based on the Bardeen Transfer Hamiltonian approach [36, 42, 43] which originates in Oppenheimer's time-dependent perturbation theory. Instead of solving the Schrödinger equation for the combined system one separates it into two independent subsystems as depicted in Fig. 3.3. The left and right electrode are described by the Hamiltonians H_L and H_R, respectively, and can be solved for by the stationary Schrödinger equation. The Hamiltonian of the full problem becomes

$$H = H_L + H_R + H_T, \tag{3.3}$$

where H_T is the Bardeen Transfer Hamiltonian which accounts for the weak coupling via the overlap of the wave functions inside the barrier region. Equation 3.3 can be read the following way: An electron in a state φ_k on the left side of the barrier feels the presence of the right electrode and the overlap as a perturbation $H_P = H_R + H_T$ which allows for tunneling transitions to states

[1] This actually serves as the proof that tunneling is the only mechanism of charge transfer in these junctions.

[2] Josephson set up the theory of Cooper pair tunneling in superconductor-insulator-superconductor junctions and the resulting effects (*dc* and *ac* Josephon effect), which also were confirmed shortly after one was able to produce the oxide tunneling barriers, but which are not subject of this thesis.

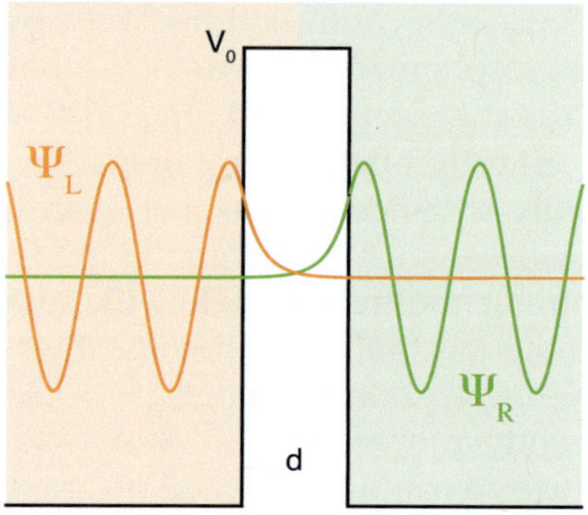

Figure 3.3: Bardeen Transfer Hamiltonian approach: In contrast to the treatment of the system as a whole in the exact description (Fig. 3.1) the wave functions of the two electrodes are described independently and the tunneling probability is determined by the overlap of the wave functions in the barrier region.

χ_q on the right. The transition rate can then be calculated from Fermi's Golden Rule [37]

$$w_{k,q} = \frac{2\pi}{\hbar} \left|M_{k,q}\right|^2 \delta(E_q - E_k). \tag{3.4}$$

Bardeen showed that the tunneling matrix element can be evaluated by integrating a current density-like operator over a surface S lying in between the two electrodes:

$$M_{k,q} = \langle \chi_q | H_P | \varphi_k \rangle = -\frac{\hbar^2}{2m} \int_S \left(\chi_q^* \vec{\nabla} \varphi_k - \varphi_k^* \vec{\nabla} \chi_q \right) \mathrm{d}\vec{S} \tag{3.5}$$

This important result allows to calculate the transition probability in a tunnel junction from the knowledge of the wave functions of the two individual electrodes only.

In order to elucidate this formula a little bit more, it is now used to calculate the transmission through the square potential barrier of Fig. 3.1 once again [44]. Due to the supposedly small transmission probability the states on both sides are described by standing waves, rather than by traveling waves, each of which is matched to a decaying function inside the barrier region the same way as before. The states on the left side are thus

$$\Psi_L = \begin{cases} \cos(kx + \delta), & x < 0 \\ g \exp(-\kappa x), & 0 < x < d, \\ 0, & x > d \end{cases} \quad \text{with } g = k/\sqrt{k^2 + \kappa^2}, \delta = \arccos(g) \tag{3.6}$$

and states Ψ_R on the right side are obtained by a similar expression with the substitution $x \to -(x - d)$ and they are taken to be standing waves for $x > d$ and 0 for $x < 0$. In this easy one-dimensional case the transition matrix element becomes

$$M \propto \left(\Psi_R^* \frac{\partial \Psi_L}{\partial x} - \Psi_L^* \frac{\partial \Psi_R}{\partial x} \right) \bigg|_a = \frac{2k^2}{k^2 + \kappa^2} \kappa \exp(-\kappa d), \tag{3.7}$$

where a is an arbitrary point inside the overlap region. After inserting this result into Fermi's Golden Rule one finds for the transition probability

$$T \propto |M|^2 \propto \frac{4k^4\kappa^2}{(k^2+\kappa^2)^2} \exp(-2\kappa d), \tag{3.8}$$

which is, apart from slightly different prefactors, the same result as found before by matching boundary conditions. Besides this educational example, Eq. 3.5 provides also a convenient starting point for calculating the matrix element in the more realistic scenario of the scanning tunneling microscope in Sec. 3.2.

The basic quantity in all tunneling experiments is the voltage-dependent tunneling current $I(U)$.[3] Unless the junction is biased, spontaneous tunneling in both directions cancels. When a voltage is applied, however, the chemical potentials of the electrodes are shifted with respect to each other and a net current will flow which can be obtained by multiplying the Golden Rule by $2e$ taking into account the charge of the electron and the two spin channels. Summing over all initial and final states one finds

$$\begin{aligned} I = \frac{4\pi e}{\hbar} \sum_{k,q} \Big(f(E_k)\,(1-f(E_q)) - f(E_q)\,(1-f(E_k)) \Big) \\ \times |M_{k,q}|^2\, \delta(E_k - (E_q - eU)) \end{aligned} \tag{3.9}$$

where $f(E_{k,q}) = (1+\exp((E_{k,q}-E_F)/k_B T))^{-1}$ is the Fermi function which describes the occupation of the states at finite temperature and accounts for the fact that tunneling can only take place from filled to empty states. Because of the continuum of states the discrete sums have to be replaced by integrals and with $D_{L,R}(E)$, the density of states of the left and right electrode, respectively, the tunneling current becomes

$$I = \frac{4\pi e}{\hbar} \int_{-\infty}^{\infty} \big(f(E-eU) - f(E)\big) D_L(E-eU) D_R(E)\, |M(E,U)|^2\, \mathrm{d}E. \tag{3.10}$$

In the case of low temperature, the Fermi function can be approximated by a step function and one obtains

$$I \propto \int_{E_F}^{E_F+eU} D_L(E-eU) D_R(E)\, |M(E,U)|^2\, \mathrm{d}E. \tag{3.11}$$

This important result of Bardeen shows that, if it is assumed that the matrix element does not change significantly in the interval of interest, the tunneling

[3] Although this thesis is written in English, the German convention of abbreviating the voltage by U instead by V is retained since it does not lead to a confusion with the unit symbol V for volt.

current is given by a convolution[4] of the densities of states of both electrodes [43] which lies at the heart of tunneling spectroscopy.

Before describing the details of tunneling spectroscopy in the last sections of this chapter, the following one presents yet another groundbreaking invention which is also based on electron tunneling: the scanning tunneling microscope.

3.2 Scanning Tunneling Microscopy

In the early years tunneling spectroscopy was restricted to junctions with a thin insulating oxide layer as barrier. Vacuum tunneling between two metal electrodes was demonstrated by Young *et al.* (1971) and by Teague (1978) [43]. Nevertheless, it were G. Binnig and H. Rohrer who, in 1982, combined the mechanism of bringing a metal tip so close to a sample that a tunneling current starts to flow with the possibility of scanning this tip over the sample surface in order to get an atomically resolved image of the latter. For their invention of the scanning tunneling microscope Binnig and Rohrer were awarded the Nobel Prize in 1987.

As already mentioned before, there is a strong dependence of the tunneling current on the width of the barrier, i. e. on the tip-sample separation. In the case of true metal-vacuum-metal tunneling the barrier height is given by the work functions of the metals, a typical value of which is $\Phi = 4\,\mathrm{eV}$. Considering Eq. 3.2 this implies that the tunneling current changes by almost one order of magnitude when the distance between tip and sample is altered by 1 Å.

By applying a voltage to piezoelectric transducers it is possible to position the tip with picometer precision. In order to obtain an image of the sample surface topography the STM is operated in the *constant current mode*. A feedback loop acts on the z-piezo and adjusts the tip-sample separation such that the measured tunneling current is equal to a reference value ("set point") which leads to an equilibrium z-position of the tip. As the latter is scanned over the surface by applying appropriate voltages on the x- and y-piezos, the z-position is readjusted by the feedback loop and thus it follows the topography of the sample and an image of the latter is obtained from the corresponding voltage at the z-piezo.

On non-homogeneous samples an STM image not only reflects the topography because, as already seen before, the tunneling current not only depends on the width of the tunneling barrier but also on the electronic density of states. J. Tersoff and D. R. Hamann applied the Bardeen model (Eq. 3.5) to the geometry of the STM modeling the apex of the tip as a sphere with s-wave symmetry

[4]To be more precise, the current (in this notation) is given by the (mirrored) cross-correlation function of the two DOS, but due to the similarity the term "convolution" is commonly used in literature.

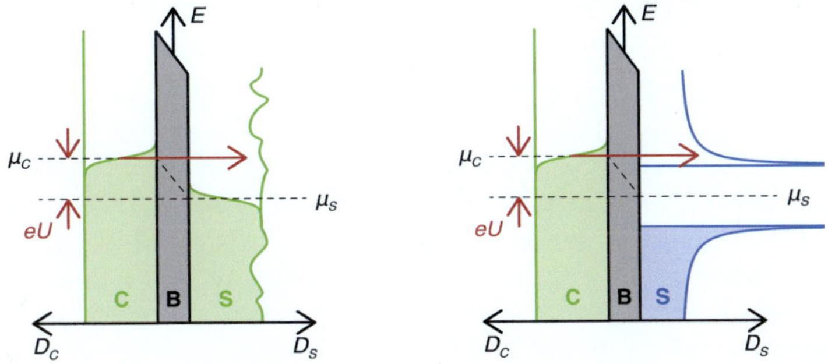

Figure 3.4: Illustration of electron tunneling spectroscopy: (a) Electrons tunnel from the left hand side normal metal electrode with constant density of states into a sample on the right. The occupation probability may be smeared at finite temperature. By ramping the voltage U the energy dependence of the DOS of the sample is probed. (b) A superconducting sample is represented by its mirrored quasiparticle DOS as a semiconductor with the energy gap 2Δ.

which yields [45]

$$I(\vec{r}_0, U) \propto D_T(E_F) \int_0^{eU} D_S(\vec{r}_0, E_F + \varepsilon) \mathrm{d}\varepsilon. \tag{3.12}$$

The tunneling current is thus proportional to D_T, the electronic density of states of the tip, which is here taken to be constant around E_F and to the integrated local density of states (LDOS) of sample, D_S, at the position of the tip. Accordingly, an STM image is always a measure for the contour of constant integrated density of states.

Since its invention the STM has become a widely used tool in surface science as well as in solid-state physics in general because it not only allows to image surfaces with atomic resolution but also to perform tunneling spectroscopy on nanostructures.

3.3 Electron Tunneling Spectroscopy

3.3.1 Probing the Density of States of Metals and Superconductors

The principle of electron tunneling spectroscopy relies on the dependence of the tunneling current on the density of states of the two electrodes in a tunnel junction. Supposed that one is interested in the density of states of one electrode S (the sample) the differential conductance $G(U)$ is obtained from the Bardeen model in the limit of low temperature (Eq. 3.11) to be

$$G(U) = \mathrm{d}I/\mathrm{d}U \propto |M|^2 D_C(E_F) D_S(E_F + eU), \tag{3.13}$$

where the matrix element and the density of states of the counter electrode C (tip) are assumed to be constant. The differential conductance is thus a direct measure of the density of states of the sample as illustrated in Fig. 3.4(a).

The importance of electron tunneling spectroscopy in the field of superconductivity arises from the fact that this picture holds in exactly the same way for quasiparticle tunneling. Even though quasiparticles are more complex objects, the superconductor can be treated in the simple semiconductor model with the energy gap 2Δ by mirroring the quasiparticle density of states (Eq. 2.9) at the chemical potential as shown in Fig. 3.4(b) [7].

If the effect of finite temperature cannot be neglected Eq. 3.10 has to be applied. Taking the other assumptions as before the differential conductance then reads

$$\mathrm{d}I/\mathrm{d}U \propto \int_{-\infty}^{\infty} D_S(E) g(E - eU) \mathrm{d}E \tag{3.14}$$

and reflects a signal which is given by the convolution[5] of the density of states of the sample with $g(x) = -\partial f(x)/\partial x$, the derivative of the Fermi function. It has roughly the shape of a Gaussian with a width of $3.4 k_B T$, limiting the energy resolution of tunneling spectroscopy.

When tunneling spectroscopy is performed in a STM setup, the feedback loop is switched off, keeping a constant distance between tip and sample while ramping the voltage. This can be done at distinct chosen positions on the sample surface after having taken an image of the latter. This kind of spectroscopy is thus often referred to as scanning tunneling spectroscopy (STS). The first successful STS measurements were published in 1986 by Feenstra *et al.* [46, 47]. A further operational mode is to record the $\mathrm{d}I/\mathrm{d}U$ signal simultaneously with the image, which results in a map of the LDOS at the preset tunneling voltage. Especially if the STM is intended to be used for tunneling spectroscopy, one has to care about the tip density of states. In Eq. 3.12 the latter was assumed to be constant around E_F which is unfortunately hard to achieve in reality. A common method is to locally melt the apex of a tungsten tip yielding a rounder shape. Such tips were shown to provide reproducible spectra [43]. In case of doubt one has to repeat the measurement with different tip conditions (e. g. after a cleaning pulse or a tip crash) in order to exclude tip effects.

In order to obtain a tunneling spectrum from the experiment, in principal, one could record the I-U characteristic and subsequently calculate the the numerical derivative. In general, however, this procedure does not yield satis-

[5] Actually, this is again the mirrored cross-correlation. If both functions are even it is however identical to the convolution.

factory results because of the noisy signal. Instead, the lock-in technique is applied, which allows to largely improve the signal-to-noise ratio. By a lock-in amplifier a small *ac* voltage $U_\Delta \cos(\omega t)$ is added to the constant tunneling bias U_0 causing the tunneling current to oscillate at the same frequency. Since the additional signal is small one can expand the signal to a Taylor series

$$I(U_0 + U_\Delta \cos(\omega t)) \approx I(U_0) + \left.\frac{\partial I}{\partial U}\right|_{U_0} U_\Delta \cos(\omega t) + \frac{1}{2} \left.\frac{\partial^2 I}{\partial U^2}\right|_{U_0} (U_\Delta \cos(\omega t))^2 \,. \qquad (3.15)$$

The component oscillating with $\cos(\omega t)$ is thus proportional to differential conductance. The lock-in amplifier now multiplies the incoming signal of the tunneling current with the reference signal of the internal oscillator and subsequently sends it through a low-pass filter. Thus, all components with different frequencies are suppressed and the derivative of the current can be extracted with high quality. A possible phase shift between the current and the reference signal arising from capacitive and inductive components of the circuitry has to be accounted for by adjusting the phase of the reference appropriately in order to get the maximal output signal. In the present study this was usually achieved by withdrawing the tip from the tunneling contact and then regulating the phase such that the output signal vanished.

A drawback of the lock-in technique is that the energy of the tunneling electrons is further smeared of due to the modulation voltage which leads again to a reduced energy resolution. If a high resolution is required, usually the modulation voltage is thus chosen small enough in order to not overcome the effect of temperature. Though the lock-in signal is proportional to the differential conductance, its absolute magnitude cannot be obtained from such a measurement but has to be reconstructed from the I-U curve. For the sake of convenience the $\mathrm{d}I/\mathrm{d}U$ spectra are thus commonly given as the lock-in signal in arbitrary units.

Besides the "1f" mode, as obvious from Eq. 3.15, there exists also the possibility to directly measure the second derivative of the tunneling current ("2f") by setting the reference frequency to 2ω since $\cos^2(\omega t) \propto \cos(2\omega t)$, which is mainly used in inelastic tunneling spectroscopy.

3.3.2 Inelastic Electron Tunneling Spectroscopy

The study of inelastic electron tunneling began in 1966 when R. C. Jaklevic and J. Lambe studied planar Al-oxide-Pb junctions applying the second derivative technique in order to see very small features in the electrical properties. They found peaks corresponding to changes in the conductance of about 1 % at certain voltages characteristic of vibrational frequencies of molecules contained in the barrier. Thus it was concluded that these features are due to inelastic

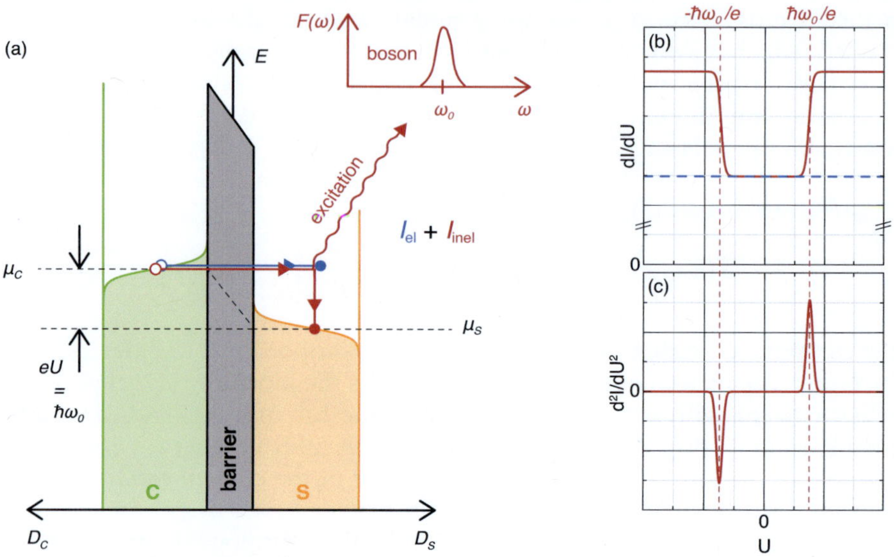

Figure 3.5: Principle of inelastic electron tunneling spectroscopy: (a) In addition to the elastic tunneling current (blue) an inelastic current flows via an additional channel (red) that opens as soon as the applied voltage is high enough in order to provide sufficient energy for exciting a boson of given energy $\hbar\omega_0$ in the sample. The conductance increases suddenly at this threshold resulting in a step in the $\mathrm{d}I/\mathrm{d}U$ spectrum (b). Due to the symmetry of electron and hole tunneling a step also appears at the same energy on the negative bias side. The characteristic signature of inelastic tunneling is thus an antisymmetric dip-peak pair in the corresponding $\mathrm{d}^2I/\mathrm{d}U^2$ spectrum (c).

electron-molecule interactions in the tunneling [48]. After realizing the spectroscopic abilities of this effect it was referred to as inelastic (electron) tunneling spectroscopy (ITS or IETS). As elastic tunneling spectroscopy, nowadays, this concept is also widely used in combination with the STM. The applications range from the vibrational spectroscopy of single molecules on surfaces [49] to the study of magnetic excitations of individual atoms or clusters [50–52]. Additionally, though rarely, there are also studies on collective excitations like spin waves (magnons) [53, 54] as well as on phonons in graphite [55]. However, no reports on the examination of phonons in superconducting materials by means of STM-IETS could be found in the literature.

The principle of IETS is sketched in Fig. 3.5(a). In direct elastic tunneling (blue) an electron tunneling from the Fermi level of the left-hand electrode arrives at a final state of energy eU above the Fermi level in the right-hand electrode which is considered as the sample of interest for instance. If there are other modes, e.g. localized vibrations or phonons of frequency ω_0, present in the sample electrons are able to interact with them and more complicated final states including the excitation play a role. This starts to happen at $eU = \hbar\omega_0$ where the electrons are just provided sufficient energy in order to end up at an unoccupied state above the Fermi level after having lost some of their energy when performing an excitation.

When the additional tunneling channel opens, the tunneling probability suddenly increases resulting in an upward step of the differential conductance at $eU = \hbar\omega_0$. As the change is fairly small, it is quite common to record not the first but the second derivative of the tunneling current with respect to the voltage[6]. In the resulting spectrum a peak appears at the voltage corresponding to the energy of the excitation [56].

If the junction is biased the other way round and electrons tunnel from right to left (sample to counter electrode) a similar argument holds, albeit tunneling takes place first and the residual hole is filled with an electron from a higher energy level under emission of an excitation. This process appears from $eU = -\hbar\omega_0$ on, also causing a step there and a dip in the second derivative. The appearance of a dip-peak pair, antisymmetric about $U = 0$, is therefore a characteristic fingerprint of an inelastic tunneling process. It is thus justified to consider the average of the positive and the negative bias side in terms of the even conductance $G_{\text{even}}(U) = 1/2(G(U)+G(-U))$ or, accordingly, the antisymmetrized second derivative of the tunneling current

$$\left(\frac{\mathrm{d}^2 I}{\mathrm{d}U^2}\right)_{\text{antisym}}(U) = \frac{1}{2}\left(\frac{\mathrm{d}^2 I}{\mathrm{d}U^2}(U) - \frac{\mathrm{d}^2 I}{\mathrm{d}U^2}(-U)\right) \tag{3.16}$$

in order to extract inelastic features [36].

As in the case of elastic tunneling spectroscopy, the energy resolution of IETS is limited by the energy distribution of the tunneling electrons due to the finite temperature and the applied lock-in technique. The measured second derivative of the tunneling current is thus given by a convolution of the intrinsic signal with the broadening functions $\chi(T)$ and $\phi(U_{\text{mod}})$ [57]. Since the shape of both functions is close to Gaussians with full widths at half maximum (FWHM) of $5.4k_BT$ and $1.7eU_{\text{mod}}$,[7] respectively, they can be replaced by one Gaussian, the FWHM of which is given by

$$\Gamma = \Delta E = \sqrt{(5.4k_BT)^2 + (1.7eU_{\text{mod}})^2}. \tag{3.17}$$

[6] In the early experiments, due to technical measurement reasons, often $\mathrm{d}^2U/\mathrm{d}I^2$ was determined instead of $\mathrm{d}^2I/\mathrm{d}U^2$. Because of the relation $\mathrm{d}^2U/\mathrm{d}I^2 = -(\mathrm{d}U/\mathrm{d}I)^3(\mathrm{d}^2I/\mathrm{d}U^2)$, where $\mathrm{d}U/\mathrm{d}I$ may be considered as a constant in the interval of interest, the spectra of both second derivatives are qualitatively comparable [56].

[7] The modulation voltages stated in this work are root mean square values ($U_{\text{mod}} = U_{\Delta,\text{rms}} = U_\Delta/\sqrt{2}$).

Accordingly, the energy resolution depends linearly on the temperature. As an example, $\Delta E = 0.37\,\mathrm{meV}$ at 800 mK without any modulation and becomes $\Delta E = 0.93\,\mathrm{meV}$ if $U_{\mathrm{mod}} = 500\,\mu\mathrm{V}$. Decreasing the modulation further, however, results in a worse signal-to-noise ratio which has to be countered by an increased measurement time. Thus, in consideration of the respective temperature, an appropriate compromise has to be found.

3.4 Boson Signatures in Tunneling Spectra

3.4.1 Phonon Signatures in the Superconducting State

Only very shortly after Giaever succeeded in experimentally accessing the energy gap in the quasiparticle density of states of superconducting Pb in a tunneling experiment [40], upon further cooling the junction below 1 K, he observed additional fine structure on the decaying edges of the quasiparticle peaks [23, 38]. His experimental result for a Pb/MgO/Mg tunnel junction is shown in Fig. 3.6(a). As explained in Sec. 2.2.3, the "wiggles" deviating from the smooth BCS shape reflect the energy-dependent gap function affected by the phonons as consistently described by Eliashberg strong-coupling theory [58]. It is therefore possible to investigate electron-phonon coupling via tunneling in a superconductor.

A semi-empirical method in order to obtain the Eliashberg function $\alpha^2F(\omega)$ and the Coulomb pseudo potential μ^* was proposed by W. L. McMillan and J. M. Rowell [9]. It is basically an inversion algorithm allowing to fit the Eliashberg function to tunneling data. Starting from a guess function for $\alpha^2F(\omega)$ and a value for μ^* chosen to yield the experimental value of Δ_0, the McMillan gap inversion algorithm solves the Eliashberg equations and compares the resulting quasiparticle density of states with the measured one. By numerically minimizing a corresponding functional derivative, iteratively the best form of $\alpha^2F(\omega)$ is found. Fig. 3.6(b) shows the Eliashberg function obtained by McMillan and Rowell applying their method to their tunneling data of a Pb/I/Pb junction.

Very often, as in the case of inelastic tunneling spectroscopy, the second derivative of the tunneling current is recorded in order to obtain clearer signals. The most prominent feature produced by a single phonon mode at ω_0 in the quasiparticle density of states of a superconductor is a strong decrease very close to $\omega_0 + \Delta_0$ (refer to Fig. 2.5). Thus, a measurement of $\mathrm{d}^2I/\mathrm{d}U^2$, when tunneling in the superconducting state of a real material with a general phonon density of states, will reveal dips on the positive bias side near its characteristic energies (van Hove singularities) shifted by the value of the energy gap and corresponding peaks on the negative bias side. The behavior is thus reversed with respect to the features expected from inelastic tunneling. Furthermore, since the strong-coupling phonon features stem from the energy dependent gap function, their intensity is expected to decrease and to vanish with superconductivity, e. g. upon heating.

The results obtained from the McMillan inversion algorithm can be used in order to calculate the critical temperature [24]

$$T_c = \frac{\Theta_D}{1.45} \exp\left[-\frac{1.04(1+\lambda)}{\lambda - \mu^*(1+0.62\lambda)}\right], \tag{3.18}$$

where Θ_D is the Debye temperature and λ is the electron-phonon coupling constant, which is defined as

$$\lambda := 2 \int \frac{d\omega}{\omega} \alpha^2 F(\omega) \tag{3.19}$$

and corresponds roughly to $D(E_F)V_0$ of the BCS theory.

Following McMillan's method tunneling spectroscopy using planar junctions was analyzed in order to study phonon-mediated superconductivity in all relevant elementary superconductors as well as in numerous superconducting alloys and wide agreement proofed that Eliashberg theory is an excellent quantitative description of conventional superconductivity [59]. However, STM experiments explicitly focusing on strong-coupling effects in conventional superconductors and the poof of the, admittedly expectable, transferability of the old concepts to the geometry of the STM are not available. Corresponding phonon signatures observed in STM tunneling spectra of lead islands were rather mentioned incidentally [60]. Anyhow, by considering complicated renormalization effects in the electronic properties of the superconducting state, this way of attaining the Eliashberg function is rather indirect.

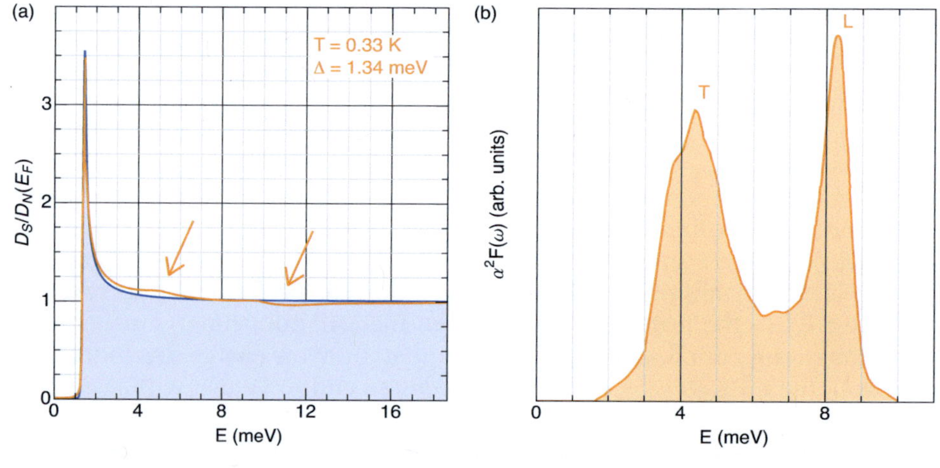

Figure 3.6: Phonon signatures in the tunneling spectrum of Pb: (a) Normalized differential conductance of a Pb/MgO/Mg tunnel junction (orange) exhibiting deviations from the BCS form (blue) marked by arrows. Data from [23]. (b) Eliashberg function of Pb extracted by the McMillan gap inversion algorithm from a similar tunneling measurement. Curve from [9].

3.4.2 Phonon Signatures in the Normal State

Since phonons and their interactions with electrons are not an exclusive property of the superconducting state, inelastic tunneling spectroscopy on metals in their normal state provides an alternative approach to the study of electron-phonon coupling. Indeed, again using planar tunnel junctions, this kind of experiments were performed around 1970 on various materials, not restricted to the ones that become superconducting in this case, but, due to the strongest electron-phonon coupling, once more it was lead which received most attention [57, 61–63]. A magnetic field, larger than the critical field of the respective material, was commonly applied in order to suppress superconductivity because the alternative, working above the transition temperature, would have resulted in a loss of energy resolution [62].

Even though, in qualitative consistency with results from neutron scattering, inelastic tunneling spectroscopy on the normal state revealed clear signals of phonons it led a shadowy existence facing the highly praised McMillan analysis of tunneling in superconductors. This is probably partially due to the fact that the quality of spectra obtained from normal metal tunneling at that time was indeed excelled by superconducting tunneling. It was therefore concluded that the first is only a cruder measure of the phonon density of states compared to the latter [62]. Furthermore, due to the composition of the planar tunnel junctions, in order to interpret the data, one had to consider effects of tunneling electrons interacting with impurities in the oxide barrier or the oxide itself. Many of the theoretical models are thus complicated by explicitly taking into account such barrier effects. A true vacuum barrier, which became feasible thanks to the STM geometry, however, allows to confine the consideration to the effects in the electrodes.

Concerning the interpretation of IETS data of planar junctions with regard to electron-phonon processes in one of the electrodes, inconsistent results were published. While many authors agree in general that a tunneling spectrum reflects the phonon density of states of the electrode of interest to some extend [57, 64, 65], there is no consensus in literature about the relationship between the $\mathrm{d}^2I/\mathrm{d}U^2$ signal and the Eliashberg function. The expression found in the textbook about tunneling spectroscopy by Wolf [36] traces back to work of Adler *et al.* [64]. This multichannel theory of inelastic electron tunneling relates the second derivative of the tunneling current to $\alpha^2F(\omega)$ but includes a kinematic factor which prevents a direct identification. Another theoretical study on inelastic processes in tunneling electrodes by Taylor [65] disagrees with some of the conclusions of Adler *et al.* Especially it does not comprise a kinematic factor but states that, under some assumptions on the momentum transfer, the differential conductance is proportional to Σ_{Im}, the imaginary part of the bulk electron-phonon self-energy in the normal state, which is related to the Eliashberg function [59].

In the light of this controversial situation, inelastic tunneling in a normal metal in presence of electron-phonon interaction was reconsidered. Based on an approach using Green's functions the elastic and inelastic contribution to the tunneling current were calculated analytically [66]. First, this calculation confirms, that phonon induced self-energy effects do not affect the elastic channel. Because the ratio of the acoustic velocity and the Fermi velocity is small, the normal state self-energy is only weakly dependent on momentum (Migdal theorem, [15]) and drops out after integrating the fermionic dispersion. Secondly, the calculation reveals that

$$\mathrm{d}^2I/\mathrm{d}U^2\Big|_{eU_0=\hbar\omega} \propto \quad G_{el}\,\alpha^2F(\omega), \tag{3.20}$$

which holds under the assumption that the Fermi surfaces as well as the Fermi velocities of both electrodes (tip and sample) are similar. This restriction arises from the fact that one of the two surface integrals in the definition of the Eliashberg function (Eq. 2.18) has to be carried out over the Fermi surface of the tip. The present calculation was performed assuming momentum conservation in the tunneling process. Since translational symmetry perpendicular to the tunneling barrier is broken, momentum in this direction, however, does not have to be conserved. In the calculation this would lead to a further averaging over momentum which might weaken the assumptions on the Fermi surfaces. $G_{el} = (4\pi e^2/\hbar)|M_{el}|^2 D_T(E_F) D_S(E_F)$ is the elastic conductance of the tunneling junction where the electronic densities of states and the matrix element are supposed to be constant around the Fermi level.

As the electronic density of states in a superconductor is strongly non-constant around E_F it becomes evident that the study of real phonon emission by inelastic tunneling spectroscopy is only possible in the normal state. Furthermore, the inelastic signal is weaker since, according to Taylor [65], the inelastic scattering coefficient is given by $|M_{kk'}|^2 = \kappa|M_{el}|^2|\alpha_{kk'}|^2$ and is thus reduced by the probability of an electron interacting with a phonon. The coefficient κ contains information about the dimensions of the electrodes and about the characteristic interaction length of electron-phonon coupling. Since, especially the latter, cannot be attained from a STM measurement, a determination of the absolute scale of $\alpha^2F(\omega)$ is not feasible.

To summarize the main result: If the electronic densities of states are flat and Fermi surface effects can be neglected, it is evident from Eq. 3.20 that inelastic tunneling spectroscopy provides a direct measure for the shape of the Eliashberg function. The second derivative of the tunneling current will thus show peaks at the energies of van Hove-singularities on the positive bias side and similar to the case of a discrete excitation the signal is point symmetric with respect to zero voltage. Since the derivative of the imaginary part of the electron-phonon self-energy is proportional to $\alpha^2F(\omega)$, the present conclusion is in agreement with the work of Taylor [65].

While the STM was introduced almost 30 years ago, still to date no experiments on the inelastic excitation of phonons in conventional superconductors are reported in literature. Even though the general advantage of the vacuum barrier ruling out any possibility of coupling of electrode phonons to those in the oxide barrier was emphasized [65] corresponding experiments were, so far, only performed on graphite [55, 67]. In this sense, there seems to be a missing link in order to properly discuss hints for bosonic modes observed in tunneling spectra of unconventional superconductors by STM.

3.4.3 Hints for Bosons in Tunneling Spectra of Unconventional Superconductors

In recent years, reports on the observation of dip-hump signatures outside the gap in tunneling spectra obtained by STM measurements on different cuprate [68, 69] and iron-based [70] unconventional superconductors were published. Though there are also attempts of explaining these signatures by band-structure effects, many authors interpret them as manifestation of a bosonic mode similar to strong-coupling effects in conventional superconductors [71, 72]. Focusing again on the iron-based materials, Fig. 3.7 shows a tunneling spectrum of $Ba_{0.6}K_{0.4}Fe_2As_2$ as an example. At energies greater than Δ the measurement clearly deviates from the provisionally assumed BCS form in a very similar way as in the case of conventional strong-coupling superconductors. Also in angle-resolved photoemission spectroscopy in a K-doping dependent study on the same material an anomalous second feature was observed [73, 74].

In contrast to conventional superconductivity, in all studies of this kind on unconventional superconductors it was supposed that the origin of the bosonic mode are spin excitations rather than phonons. This hypothesis is based on the consistence of the mode energy extracted from the tunneling spectrum with the energy of spin excitation measured by inelastic neutron scattering. Additionally, the variation of the mode energy and intensity with the local superconducting gap was regarded as strong hint that this mode is closely related to superconductivity [69, 71, 72].

The observation of fine structure in tunneling spectra in these recent experiments and its interpretation as spin excitations is already very interesting with respect to the identification of the pairing glue in unconventional superconductors. However, an examination of these effects in the more direct way as excitations in the normal state, which has not been reported in literature yet, would certainly be able to provide further insight. Driving an unconventional superconductor into the normal state by the application of a magnetic field, however, is unfortunately impossible since the corresponding values of B_{c2} are inaccessibly high. For example, $B_{c2} = 43T$ for $Ba(Fe_{0.9}Co_{0.1})_2As_2$ [75]. The alternative of performing the experiments above the respective critical tempera-

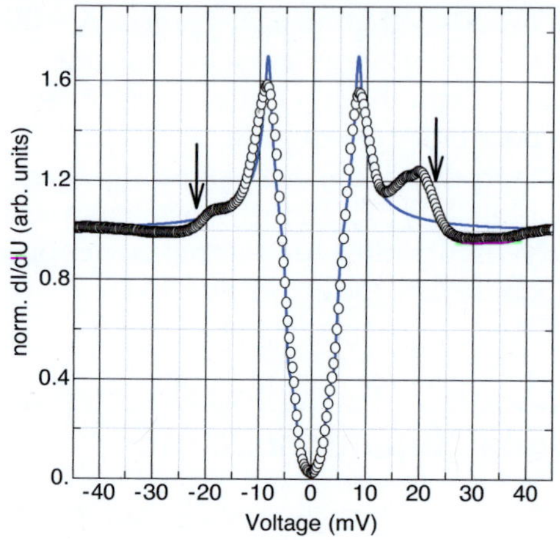

Figure 3.7: Differential conductance of $Ba_{0.6}K_{0.4}Fe_2As_2$. Outside of the energy gap the spectrum exhibits a hump-dip structure around $\pm 22\,\mathrm{mV}$ deviating from a provisional BCS fit (blue line). Adapted from [71].

ture, occasionally very high, might not provide the necessary energy resolution. Since unconventional superconductors are usually of type-II, a loophole might consist in considering the normal state in the center of a vortex, which can be approached thanks to the unique capabilities of the STM. To give an indication of the required magnetic field, B_{c1} is less than $100\,\mathrm{mT}$ for $Ba_{0.6}K_{0.4}Fe_2As_2$ [76], which is relatively small and easily accessible.

4 Experimental Setup and Methods

The experiments of this work were all performed with a Joule-Thomson low-temperature STM (JT-STM) which was built at Physikalisches Institut in the group of W. Wulfhekel and described in detail by L. Zhang [77, 78]. Together with the technical requirements on the experimental setup this chapter gives an overview of its actual realization and, furthermore, it explains the equipment for and methods of sample and tip preparation.

4.1 Requirements on the Experimental Setup

4.1.1 Setup and Control of a STM

As already explained in Sec. 3.2 the tunneling current is sensitive to changes in the tip sample separation of the order of one-hundredth of a nanometer. That is why the distance, which is typically around 1 nm, has to be stabilized with a precision of less than 1 %. This is achieved by a multi-stage damping system. First, the whole setup is supported by active air-damping system. Secondly, the STM body itself, which is very rigid providing a high resonance frequency, hangs freely on springs. Even though additional eddy-current damping using permanent magnets is prohibited by the intention of performing experiments on the magnetic field dependence, the STM is effectively isolated from vibrations of the building and the variation of the tip-sample separation is reduced to a few picometers. Under these conditions it is possible not only to image single adsorbates on a substrate but also the atomic structure of a single crystal surface itself.

Of course, one has to be able to control the position of the tip with angstom precision above the surface. This is accomplished by a single piezoelectric cylindrical tube scanner allowing the motion of the tip in all three degrees of freedom. The necessary voltages are supplied by a *Nanonis* STM controller which is operated via a corresponding computer program. The latter provides all the functionality for taking topographic images and performing tunneling spectroscopy at chosen positions. Further, the STM includes additional piezoelectric slip-stick motors allowing a coarse motion in the millimeter to sub-micrometer range of the tip in z-direction, especially needed in order to bring the tip close to tunneling contact after inserting a new sample. A first rough ap-

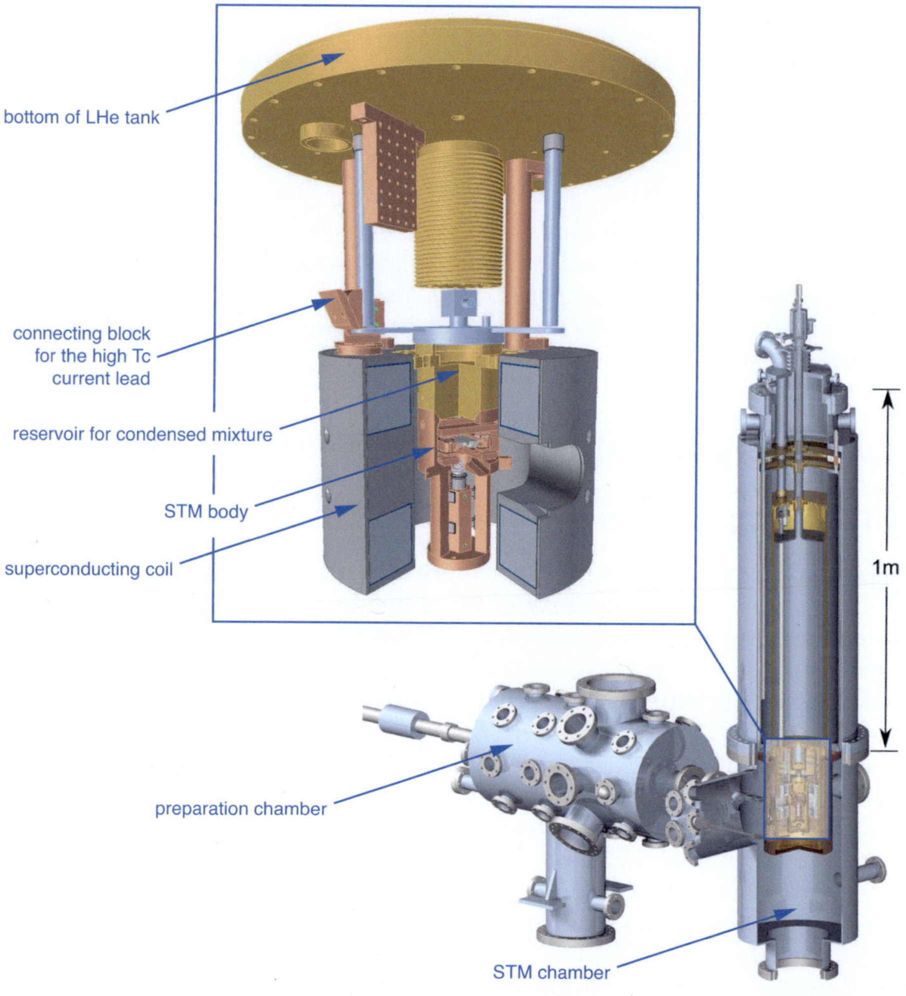

Figure 4.1: Technical drawing of the experimental setup: The full view [77] shows the preparation chamber and the STM (measurement) chamber. The inset magnifies the STM below the reservoir for the condensed mixture. The latter is mounted with hardly any thermal connection to the bottom of the LHe bath cryostat via stainless steel joints. The superconducting coil and its connections were redesigned as a part of the present work. Springs, wires and capillaries are omitted in the drawing for clarity.

proach is controlled by checking the capacitance between tip and sample since no appropriate optical access is possible.

Finally, another slip-stick motor allows to change the sample position laterally, for example if the area accessible by the tube scanner (typically around $1\,\mu\mathrm{m}^2$) should be changed during the experiment. A view of the STM in its immediate surrounding is depicted in the magnified inset of Fig. 4.1 showing a technical drawing of the experimental setup.

The resistance of a STM contact is usually of the order of mega- to gigaohms. This leads to rather small currents of about $10\,\mathrm{pA}$ to $10\,\mathrm{nA}$ when a typical voltage of several millivolts up to a few volts is applied. In order to measure the small currents a *Femto* transimpedance amplifier (current to voltage converter) with a variable gain of $10^2\,\mathrm{V/A}$ to $10^{10}\,\mathrm{V/A}$ is used that converts the current into a conveniently measurable voltage. It inherently acts as a low-pass filter with the bandwidth depending on the amplification. When performing tunneling spectroscopy, this has to be considered when setting the modulation frequency at the lock-in amplifier.

The voltage interval of interest for examining the energy gap in the superconducting quasiparticle density of state as well as the boson signatures is a few millivolts around zero. In order to reduce the influence of electrical noise at these small voltages, a 1/100 voltage divider is inserted regularly in the circuitry.

4.1.2 Ultra-High Vacuum

In order to obtain meaningful results on the intrinsic properties of a sample from tunneling spectroscopy it has to be conducted ideally on its clean, contaminant-free surface. Therefore the preparation and execution of all experiments have to take place under ultra-high vacuum (UHV). This becomes clear immediately from the fact that, at ambient conditions, it only takes $3.6 \times 10^{-6}\,\mathrm{s}$ until the sample surface is covered by one monolayer of adsorbates. Only if a pressure below $10^{-9}\,\mathrm{mbar}$ is reached this time is extended to more than one hour [79]. Thus, only in UHV there is enough time for the preparation and the examination since the latter may regularly last about one week.

To achieve the desired conditions all the experimental devices are placed inside stainless steel chambers which are evacuated by different pumps. A rough vacuum is set up by a rotary pump which also provides the pre-vacuum necessary for the operation of second step turbo molecular pumps (TMP). Due to their high pumping speed, they are used to evacuate a chamber after it was vented, but also during preparation steps where larger amounts of gas have to be pumped. Since turbo molecular pumps cause mechanical noise at their turning frequency of about $1\,\mathrm{kHz}$ they severely disturb STM measurements and are therefore turned off during the experiments. The UHV is then maintained by ion getter pumps which do not possess any moving parts. Additionally, a titanium sublimation pump can be applied to further improve the pressure. A base pressure of around 1×10^{-10} mbar is reached in the used setup, where Bayard-Alpert hot-cathode ionization gauges are used to monitor the pressure [80].

As depicted in Fig. 4.1, the setup is divided into a STM chamber and a preparation chamber, in order to avoid the necessity of venting the whole system in case of repair which has to be followed by a time-consuming bake-out procedure. Further, the sub-division allows sample preparation without affecting the pressure in the STM chamber. This is especially necessary because in the STM chamber the large, extremely cold surfaces of the cryostat to be described below act as a very efficient cryogenic pump. An additional load-lock chamber for easy insertion of new samples is installed as well. Inside the vacuum system, the transfer of samples, tips, and filaments, all mounted on standardized 15x15 mm plates, is carried out by transfer rods and mechanical hands. Further, there are storage facilities to keep several samples at a time under UHV conditions.

4.1.3 Low Temperatures

Besides the fact that superconductivity can only be studied below the critical temperature of the respective sample, low temperatures are also required to provide the necessary energy resolution for resolving boson signatures in tunneling spectroscopy. As they may appear well below 10 mV, temperatures below 1 K are needed (see Sec. 3.3.2). The JT-STM provides a base temperature below 800 mK which is achieved by a Joule-Thomson stage. A liquid helium bath cryostat, which is surrounded by a tank filled with liquid nitrogen in order to increase the standing time for helium up to more than one week, is used to reach 4.2 K. With only minute thermal connection to the helium tank, a reservoir of 10 ml is installed, which is connected to a closed cycle through which a mixture of ^{4}He and ^{3}He is pumped. Right in front of the reservoir there is a thin capillary of an inner diameter of 90 μm [77] separating the high from the low pressure side. As a liquid, the mixture reaches the small reservoir where the reduction of the vapor pressure below 1 mbar results in a boiling temperature around 800 mK [81]. The STM body is thermally anchored directly at this reservoir pot. Temperature control is accomplished by measuring the resistance of a commercial ceramic zirconium oxynitride (CERNOX) [81] temperature sensor mounted at the STM body. A time-consuming difficulty of low temperature STM experiments is the fact that one has to wait until the sample has reached equilibrium base temperature which takes several hours. The thermal contraction during cool down results in strong thermal drift preventing any STM measurements apart from very crude first checks.

4.1.4 Superconducting Coil

In order to suppress superconductivity, at least partially in the case of unconventional superconductors, high magnetic fields of several tesla are desired. Concerning the low-temperature setup, such fields have to be produced necessarily by a superconducting coil. However, commercial coils, produced by standard procedures, are not compatible with UHV conditions and therefore custom-made coils using appropriate materials have to be employed. Mainly aiming at the enhancement of the maximal field, a redesign of a previously installed coil, which was only able to reach 1 T, and the fabrication of a new coil was part of the present work.

Among the requirements on the correct operation of a superconducting coil are its mechanical stability in order to structurally withstand large magnetic stress and its thermal stability. The JT-STM setup requires a split-coil design which allows to access the STM placed near the geometrical center of the coil. The force between the two winding sections separated by 35 mm corresponds to several tons of weight. The necessary mechanical stability is achieved by a bobbin which is made of massive aluminum filling the whole volume in between the two sides except for an access opening in front. Thermal stability is very important since, when parts of the superconducting wire become resistive, the magnet should be able to tolerate the resulting heat to some extent in order to prevent a "quench", the process of a magnet totally losing superconductivity [82]. However, the necessary cooling is hard to achieve in a dry-magnet design where it is not put directly into liquid helium but mounted inside UHV. Therefore, the installation requires massive copper connections between the coil and the helium tank (see Fig. 4.1). Furthermore, the bobbin as well as the glue used to fill the remaining space between the windings have to be good thermal conductors.

The risk of "quenching" a superconducting magnet can be minimized by generally designing it to operate at a current which is only 5070 % of its maximum possible current [83]. Therefore, in the redesign, a thicker NbTi wire (Supercon54S43, [84]) with a bare diameter of 0.3 mm was used which provides higher stability due to a larger ratio of the copper matrix to superconducting filaments (1.3/1) and a higher critical current. Its dependence on the magnetic field is illustrated in Fig. 4.2(a). A "quench" will be initiated at locations of strongest local magnetic field, since there the critical current of the wire is reached first. Figure 4.2(b) shows a finite-elements simulation of the magnetic field distribution of the redesigned magnet, which also possesses increased dimensions thus optimizing the usage of the available space as far as possible. The ratio of the magnetic fields at the position of the sample close to the center and at the position of maximum field at the inner diameter of the winding sections, as marked in Fig. 4.2(b), is 68 %. The corresponding load lines are given in Fig. 4.2(a). One finds the maximal possible current from the

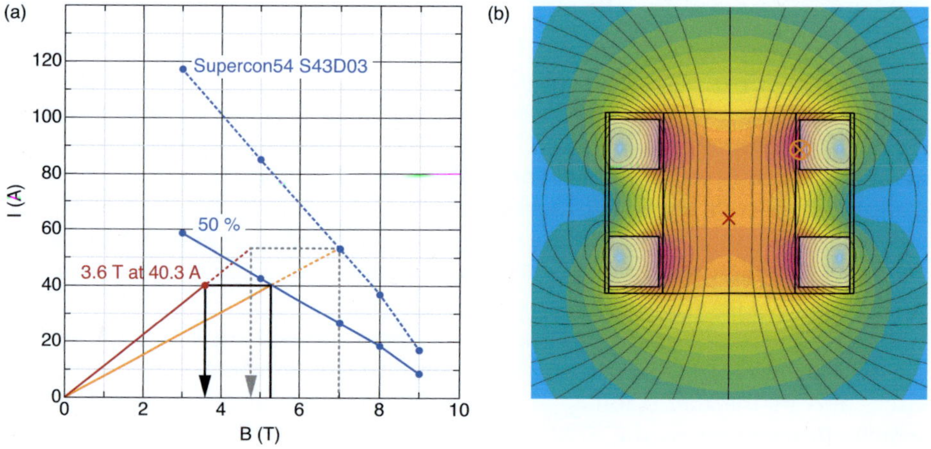

Figure 4.2: Redesign of the superconducting magnet: (a) Critical current $I_c(B)$ of the used NbTi wire (blue, dashed) [85] and its 50 % safety values (blue, straight) together with the load lines at the position of the sample (red) and at the position of maximum field (orange) as marked in (b), a finite-elements simulation of the magnetic field distribution. The winding sections are highlighted in the superimposed technical drawing.

interception of the load line at the position of maximum field with the $I_c(B)$ curve of the wire. Considering the load line at the position of the sample, the maximum field would be 4.7 T at $I_{\mathrm{max}} = 53.3\,\mathrm{A}$. However, keeping the safety margin of 50 %, one finds 3.6 T at 40.3 A.

The new NbTi wire has a diameter of $0.34\,\mathrm{mm}$ including the insulation [85]. Compared to the previous coil, this resulted in less windings. In the manufacturing process, around 5600 windings on either side could be achieved corresponding to a fill factor of 77 %, the ratio of the cross-section of all windings to the total cross-section[1]. For stabilization against conductor motion the void in between the windings was filled by UHV compatible, thermal conductive Epotek H77S glue. However, the winding sections are not glued to the coil form itself, but are "floating" on separating teflon sheets ([83], p. 412).

Since there are less winding in the new coil, higher load currents than before are needed. In order to afford the increased currents, it was necessary to redesign the electrical input lead. This is made from a high-temperature superconductor material to minimize additional heat conduction and thus to keep

[1]The fill factor of a hexagonal close-packed winding would be 90.7 %. In the present coil this ordered configuration is only maintained during the first few layers. The rest was accomplished in the so-called "wild winding". In total, the achieved fill factor is very close to 78.5 %, the one of cubic close-packed configuration.

the long standing time of liquid helium, which could not be achieved using a standard copper lead.

After the installation of the new coil in the experimental setup it was tested up to fields of 3 T, which can now be reliably applied for the experiments. The bottleneck, at present, is the high-temperature superconducting current lead, which has to be accounted for by reducing the temperature of the liquid nitrogen reservoir below 77 K using an additional pump.

4.2 Facilities for *in situ* Sample Preparation

4.2.1 Sputtering and Annealing

Contamination free surfaces of single crystal samples are best suited for STM measurements. Commonly, they are obtained by a standard procedure consisting of Ar^+ sputtering followed by annealing at high temperature [79]. The preparation chamber is equipped with all devices and facilities needed to clean samples and to check their surface quality. A sputter gun allows to shoot Ar^+ ions with a kinetic energy of more than 1 kV onto the sample under an incidental angle of $45\,^\circ$ with respect to the sample surface thus removing the topmost layers within a time of around half an hour. The resulting corrugation is subsequently smoothened by an annealing step where the sample is heated from the back by electron bombardment. For this purpose, the head of the manipulator provides a second slot carrying a filament which provides thermal electrons. By applying a voltage between the filament and the sample of up to 1000 V, the electrons are accelerated towards the sample. The heating process is controlled by various single band or more advanced two color ratio pyrometers, depending on the temperature range of interest. While in case of samples that were just exposed to air, several cleaning cycles are necessary, only one may be enough if the sample was stored in vacuum. In order to examine the quality of the sample surface the preparation chamber provides a combined LEED and AUGER device allowing to unravel the crystal structure of a surface by low energy electron diffraction (LEED) and its chemical composition by Auger electron spectroscopy, respectively.

The performance of the sputtering and annealing procedure strongly depends on the nature of the sample. While satisfactory results are quite easily obtained for copper, a clean surface of niobium is much harder to achieve (see Sec. 5.2). Since annealing is not able to recover flat surfaces after sputtering silicon, a purely thermal treatment has to be chosen in this case. Due to the complicated crystal structure of unconventional superconducting materials the method introduced before completely fails and clean surfaces have to be obtained from *in situ* sample cleavage, details of which will be described in Sec. 7.2.

4.2.2 Molecular Beam Epitaxy

Lead islands on copper were also fabricated in the preparation chamber by molecular beam epitaxy (MBE) from an evaporation source in which the metal to be deposited is heated by electron bombardment from a concentric tungsten filament. During evaporation a certain gas density is established by sublimation. A specific ratio of the atoms is ionized and the corresponding current to a flux electrode can be used to calibrate the deposition rate [86]. The latter can be adjusted by setting the filament current and the high voltage accordingly and is of the order of one atomic layer per minute. Thanks to water cooling of the surrounding parts in the source, the base pressure in the chamber is almost unaffected during operation of the MBE source guaranteeing very pure structures on the sample.

4.2.3 Preparation of STM Tips

In all experiments an electrochemically etched tungsten wire is used as STM tip. After insertion into UHV, tips are generally treated by several cycles of Ar^+ sputtering and flash-annealing by electron bombardment at 1 kV and emission currents up to 50 mA. This procedure removes contaminations and oxides from the freshly etched tip. The flash-annealing also results in local melting of the tip end which then has a larger radius after recrystallization. Such tips were empirically found to be well suited for tunneling spectroscopy [43]. However, when imaging sharp and high steps, for example the border of a three dimensional islands, a rather wide tip apex results in multiple tip features and may complicate the interpretation of topographical images, especially in case of a high island density. If necessary, tip treatment can be repeated *in situ* since the JT-STM setup allows easy tip exchange and trasnfer.

5 Niobium

After the theoretical background and the idea for the experimental approach to the electron-phonon coupling were discussed in the previous chapters, now the STM study of a niobium single crystal is presented. First, a motivation for the choice of this system is given by summarizing existing results. The necessary preparation steps in order to obtain a surface quality appropriate for a STM study are explained before the results regarding phonon signatures in tunneling spectra of the superconducting as well as of the normal state are presented. Since the latter was achieved by applying a magnetic field, its effects on the sample are discussed in some detail.

5.1 Motivation

As can be seen from Tab. 5.1, which lists the relevant properties of niobium, the transition temperature is $T_C = 9.2\,\mathrm{K}$ being the highest value among the elemental superconductors. The pairing mechanism is by conventional phonon-mediated coupling. The superconducting energy gap is $\Delta_0 = 1.45\,\mathrm{eV}$ resulting in a ratio of $2\Delta_0/k_BT_C = 3.66$, which is only a little higher than the BCS value of 3.53 and thus niobium is among the weak to intermediate coupling superconductors.

Nevertheless it is possible to observe fine structure in the tunneling conductance of superconducting niobium reflecting the electron-phonon coupling. Figure 5.1(a) shows the result of Bostock *et al.* [90] on a planar Nb-NbO-In tunnel junction. In agreement with Eliashberg theory dips are observed at around $18\,\mathrm{meV}$ and $25.5\,\mathrm{meV}$. These dips are thus very close to the energy of van Hove singularities in the phonon density of states at $15.5\,\mathrm{meV}$ and $24\,\mathrm{meV}$ stemming mainly from the transversal and longitudinal phonon modes, respectively, shifted by the sum-gap value $\Delta_{0,Nb} + \Delta_{0,In} \approx 2\,\mathrm{meV}$. From this experiment, using McMillan's inversion algorithm, the authors subsequently extracted the Eliashberg function presented in Fig. 5.1(b). The accordance with the phonon density of states from neutron-scattering [91] is striking. Only the height of the longitudinal peak is smaller, probably indicating a reduced electron coupling to longitudinal modes.

Although the shape of the Eliashberg function obtained by Bostock *et al.* clearly is accurate, the extracted values for λ and μ^* deviate substantially from the expectation, especially μ^* which is negative. Before, in 1971, Shen, who was the first to report Nb tunneling data [92], found the same inconsistency. It was

property	value	reference
T_c	9.2 K	[7]
Δ_0	1.45 meV (1.56 meV)	[7] ([87])
$2\Delta_0/k_BT_c$	3.66	[7]
ξ_{GL}	106 nm	[7]
λ_L	85 nm	[7]
κ	0.8	[7]
B_{c1}	184 mT at 0.8 K	[88]
B_{c2}	376 mT at 0.8 K	[88]
crystal structure	bcc	[89]
lattice constant	330 pm	[89]

Table 5.1: Summary of relevant physical properties of Nb.

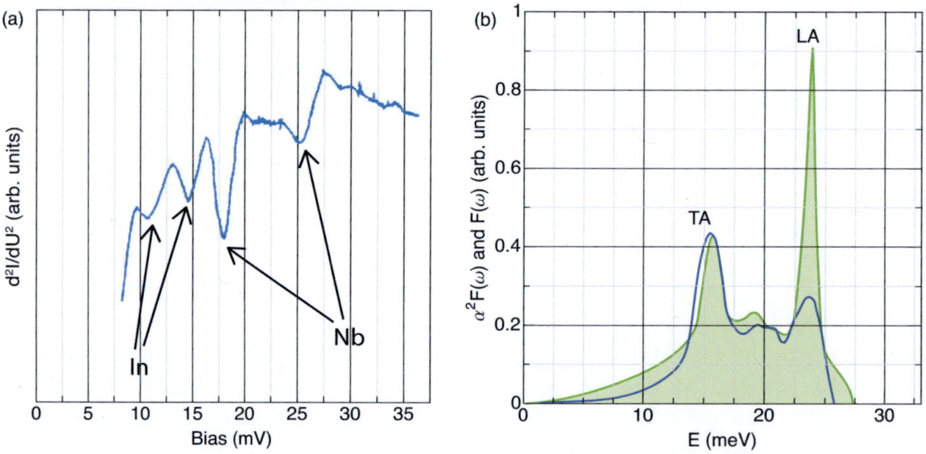

Figure 5.1: Phonons in niobium: (a) $\mathrm{d}^2I/\mathrm{d}U^2$ spectrum of a planar In-NbO-Nb tunnel junction with Nb and In in the superconducting state. (b) Phonon DOS (green) obtained by neutron scattering [91] and Eliashberg function extracted from (a) using McMillan inversion routines (all curves adapted from [90]).

suggested in this case that the disagreement originates from contaminations at the niobium-insulator interface. In fact, the purity and structural perfection of the Nb foil used in this study was not of the same high quality as the Nb samples employed by Bostock *et al.* Later, Wolf *et al.* argued that data obtained from tunnel junctions comprising niobium oxide are in general not suited for extracting physically meaningful values for λ and μ^* [93]. Instead they proposed a junction type involving an additional normal metal layer preventing the oxidation of niobium. Since the tunneling does not take place directly in the superconductor but in the normal metal, this technique is referred to as proximity electron tunneling spectroscopy (PETS) [36, 93]. Due to the extended experimental system an even more complicated theory is necessarily employed in order to reconstruct $\alpha^2 F(\omega)$, λ and μ^*, which in turn gave rise to doubts [87]. The purpose of the present work, in the first instance, is to investigate whether it is possible at all to detect the phonon modes of a material in the superconducting and in the normal state in the geometry of the STM. Therefore the knowledge of the characteristic energies is sufficient at the moment.

All the studies cited above were carried out on superconducting niobium. It seems, that tunneling spectroscopy has not been performed yet on Nb in the normal state. At least no such tunneling experiments were found in literature. One way of driving niobium to the normal state would be to increase the temperature above T_c. Since this results in a loss of energy resolution in the tunneling spectra (see Eq. 3.17), another way shall be chosen here, namely the application of a magnetic field.

According to Ginsburg-Landau theory, the response of a superconductor to the application of an external magnetic field is determined by κ, the Ginsburg-Landau parameter, which gives the ratio of two fundamental properties, namely the London penetration depth λ_L and the coherence length ξ_{GL} [7]. If this ratio is less than $1/\sqrt{2}$, the superconductor is of type-I and completely expels a magnetic field as long as it is smaller than a temperature-dependent critical field $B_c(T)$ (Meißner phase) and becomes normal conductive for larger fields (normal phase). If, however, $\kappa > 1/\sqrt{2}$ one deals with type-II superconductivity which is characterized by an additional intermediate phase in the thermodynamic B-T diagram. In this so-called Shubnikov phase between a lower critical field B_{c1} and an upper critical field B_{c2} the magnetic flux passes through the material in form of Abrikosov vortices possessing a normal conductive core screened by circulating supercurrents. Each vortex thus carries a fraction of the external magnetic flux quantized in units of $\Phi_0 = h/2e$, the magnetic flux quantum. In clean type-II superconductors without pinning centers the vortices can freely move and therefore arrange in an ordered Abrikosov lattice, which is usually hexagonal. The vortex density is proportional to the

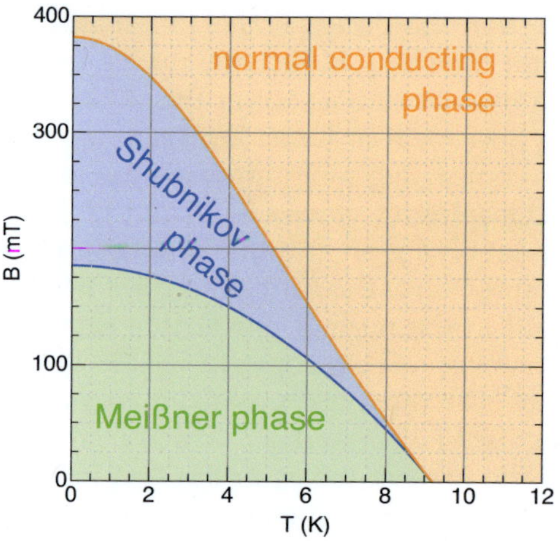

Figure 5.2: Phase diagram of niobium: Since Nb is a type-II superconductor there is the intermediate Shubnikov phase in between the purely superconducting Meiner phase and the normal conducting phase. The separation lines were calculated from the formulas given in [88].

external magnetic field B and the distance between two vortices is given by [7]

$$d = \sqrt{\frac{2\Phi_0}{\sqrt{3}B}} = 1.075\sqrt{\frac{\Phi_0}{B}}. \tag{5.1}$$

As expected from the Ginsburg-Landau parameter $\kappa = 0.8 > 1/\sqrt{2}$, niobium is a type-II superconductor. The lower and upper critical fields, $B_{c1} = 184\,\mathrm{mT}$ and $B_{c2} = 376\,\mathrm{mT}$ at the experimental temperature $T = 0.8\,\mathrm{K}$, can be read from the phase diagram given in Fig. 5.2. This phase diagram was calculated from the following formulas valid for niobium of high purity [88]:

$$B_{c1}\,[T] = 0.185\left(1 - \left(\frac{T}{T_c}\right)^2\right) \tag{5.2}$$

$$B_{c2}\,[T] = 0.382\left(\frac{1 - (T/T_c)^2}{1 + (T/T_c)^2}\right), \tag{5.3}$$

with $T_c = 9.2\,\mathrm{K}$. With the newly constructed magnetic coil the necessary fields can easily be reached and tunneling spectroscopy in the normal state can thus be performed either in the fully normal conducting phase or in the center of a vortex in the intermediate Shubnikov phase, though, in the latter case, the appearance of bound states [94] might prevent an easy identification of the relevant phonon features.

5.2 Preparation of the Nb(110) Surface

In this work the (110) surface of a Nb single crystal with a purity of 99.999 % (MaTeck GmbH) was used as a sample for the examination of the electron-phonon interaction. The same crystal had been already studied in the *PhD* thesis of T. Tomanic in a different context [95]. After mechanical and electrochemical polishing the cylindric crystal, which had a diameter of 5 mm and a height of 1 mm, it was fixed by spot welded tantalum cramps on a molybdenum plate. A hole with a diameter of 4.5 mm had been drilled into the sample plate to allow a direct heat treatment from the backside as well as to reduce the thermal contact.

The Nb(110) surface was widely studied in the past [96–100]. Since it is very difficult to obtain a really clean Nb(110) surface, however, there are only few studies on absolutely pure surfaces. For example, An *et al.* were able to examine the influence of oxidation of such a pure surface [97]. After a Nb(110) crystal was exposed to air the main contaminants are carbon, nitrogen and oxygen with the latter building a complex layer of various strongly bound Nb oxides (NbO, NbO_2, Nb_2O_5 [101–103]). Especially along grain boundaries oxygen also diffuses into the Nb single crystal and gets dissolved there. While carbon and nitrogen can be removed rather easily by Ar^+ sputtering for several hours [99] it is much more difficult to eliminate oxygen contaminations from the surface since oxygen is continuously segregating from the bulk. In order to study the electron-phonon coupling in niobium by tunneling spectroscopy, however, a clean surface is very much desirable since remaining oxides could corrupt the spectra of pure niobium. Furthermore, it is known that a minor content of oxygen dissolved in the niobium crystal generates a distinct decrease of the transition temperature [104] as well as of the energy gap [87]. Clean surfaces could only be obtained by fast heating (flashing) the crystal to temperatures above 2000 °C followed by a rapid cool-down [95, 97].

The preparation procedure applied in this study is essentially based on the one proposed by Tomanic [95]. After sputtering the Nb(110) surface for 15 min to 30 min with Ar^+ ions (3 kV, 3 μA) the sample was flashed about ten times to 1500 °C to 1800 °C by electron bombardment (1 kV, 110-140 mA). The temperature was controlled using a two color ratio pyrometer. This procedure was repeated several times and the quality of the surface was checked by STM from time to time. While already after the first few cycles apparently flat terraces with a width of about 40 nm could be observed in a large scale STM image (Fig. 5.3(a)) the closeup presented in Fig. 5.3(b) reveals that the surface was still covered by well-known NbO chains consisting of 10±1 Nb atoms [95]. In total, the crystal was sputtered for more than 9 h including about 3 h of hot sputtering at 500 °C and flashed more than 150 times. The STM image in Fig. 5.3(c) shows the resulting topography of the surface at the atomic scale. This is still not the pure Nb(110) surface but a superstructure appearing in different domain

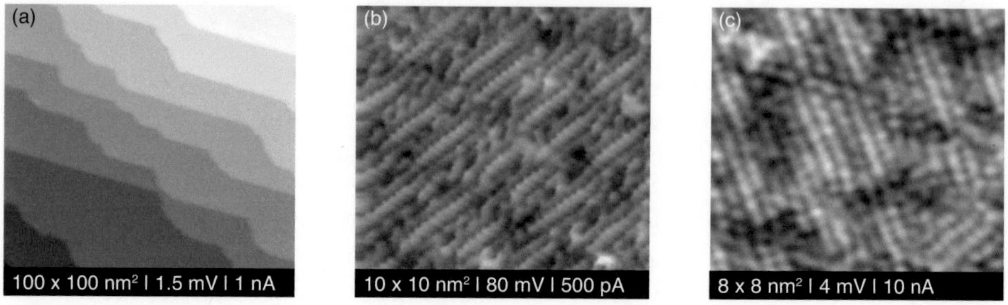

Figure 5.3: STM images showing the topography of the Nb(110) surface: (a) Apparently flat terraces with a width of about 40 nm. (b) Atomic scale image recorded after a few preparation cycles revealing NbO chains remaining on the surface. (c) Different surface topography present in some areas after more preparation cycles.

orientations which was also observed most of the time by Tomanic and could arise from carbon impurities segregated to the surface as speculated there [95]. It has to be mentioned that also on this kind of surface there were still NbO chains present on some areas. Since the bare Nb(110) surface was not obtained in the present study it might be argued that the number of cleaning cycles was not sufficient but on the other hand even Tomanic, who performed about 100 cycles, was able to observe the bare Nb(110) surface only very rarely.

For this reason it was decided to check the quality of the sample surface at this stage by tunneling spectroscopy. Fig. 5.4 shows the obtained spectrum clearly revealing the superconducting density of states. Almost perfect agreement is obtained by fitting to the data the BCS gap function (Eq. 2.9) convolved with the derivative of the Fermi function reflecting the energy distribution of the electrons in the normal conducting tip at the finite measurement temperature. Since the convolution cannot be calculated analytically, a *Mathematica* code was written capable of performing the fit numerically. Besides the energy gap Δ and the temperature T the model contains four more free parameters accounting for eventual offsets in the voltage and the conductance, the normalization and a possible overall tilt of the spectrum. Not only the shape of the spectrum but also the extracted value of $\Delta = 1.52\,\mathrm{meV}$ lies well in the range of literature values for pure niobium [7, 90]. Only the temperature resulting from the fit is with $T_{\mathrm{fit}} = 1.4\,\mathrm{K}$ substantially higher than the one determined by the Cernox thermometer ($0.85\,\mathrm{K}$). On the one hand, this discrepancy could be partially attributed to the additional broadening due to the lock-in technique as well as unknown electronic noise. On the other hand it might well be that the temperature of the electron system is in fact not as low as the thermometer suggests. All in all, it can be concluded that the quality of the obtained surface should be suited well enough in order to study the electron-phonon interaction.

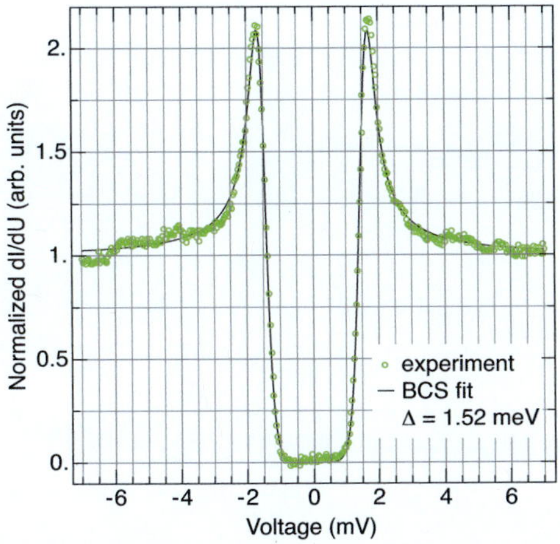

Figure 5.4: Superconducting gap of niobium: The $\mathrm{d}I/\mathrm{d}U$ spectrum measured at $0.85\,\mathrm{K}$ using lock-in technique ($U_{mod} = 100\mu V$) is in perfect agreement with the convolved BCS DOS. The resulting energy gap is in the range of literature values [7, 90].

5.3 Experimental Results

Phonon features in the superconducting state

Regarding the study of the electron-phonon interaction the first step consisted in checking whether it is possible with the STM setup to reproduce the results obtained with planar tunnel junctions, namely the observation of phonon signatures altering the superconducting density of states via Eliashberg strong coupling theory. Therefore, in the beginning, a $\mathrm{d}^2I/\mathrm{d}U^2$ spectrum was recorded in the superconducting state (Fig. 5.5(a)). As expected, at energies close to zero this spectrum is dominated by a feature reflecting the derivative of the superconducting energy gap while there is nothing to recognize at voltages above $\pm 15\,\mathrm{mV}$. Only after magnification of this spectrum by a factor of 100 one can clearly identify two peak-dip pairs at $\pm 17\,\mathrm{mV}$ and at $\pm 25.5\,\mathrm{mV}$ in 5.5(b). These positions correspond exactly to the energies of the "transversal" and "longitudinal" van Hove singularities in the phonon density of states of Nb shifted by $\Delta_{\mathrm{Nb}} = 1.5\,\mathrm{meV}$. This measurement thus is in agreement with the previous results on planar Nb tunnel junctions and once again shows experimentally that, in the superconducting state, shifted dips appear at the positive bias side of the $\mathrm{d}^2I/\mathrm{d}U^2$ spectrum while peaks right at the phonon energies are expected in the picture of IETS. In order to see whether the phonons can also be detected as inelastic excitations by the tunneling electrons the Nb sample has therefore be forced to pass into its normal state by applying a magnetic field.

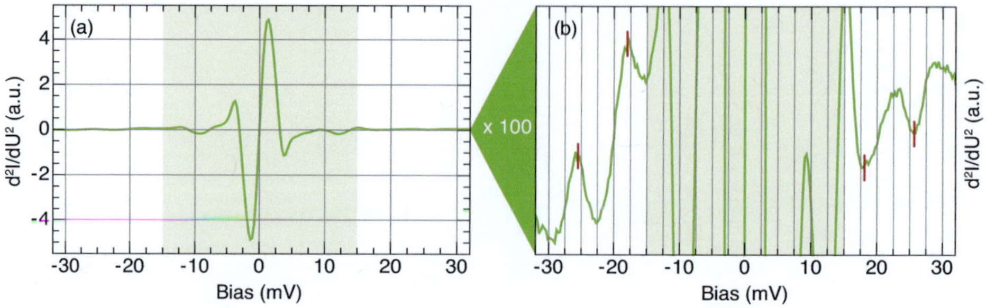

Figure 5.5: $\mathrm{d}^2I/\mathrm{d}U^2$ tunneling spectrum of superconducting niobium: (a) Lock-in signal (b) Magnification of (a) by a factor of 100. Outside the marked region which is dominated by the signature of the superconducting gap two clear peak-dip pairs show up at $\pm 17\,\mathrm{mV}$ and at $\pm 25.5\,\mathrm{mV}$.

Phonon features in the normal state

Fig. 5.6(a) shows the average of several $\mathrm{d}^2I/\mathrm{d}U^2$ spectra recorded at $0.8\,\mathrm{K}$ when a magnetic field of $B = 1.6\,\mathrm{T}$ perpendicular to the surface was applied. This high field, well above B_{c2} of pure niobium, was necessary because it is known that impurities in the crystal lead to an enhanced upper critical field [105]. In fact, for the studied crystal one could safely assume the normal state from STS spectra only at fields above $1\,\mathrm{T}$.

In contrast to the peak-dip pairs in the spectrum from the superconducting state, now the dips at $17\,\mathrm{mV}$ and $25.5\,\mathrm{mV}$ at the positive bias side turned into peaks at slightly shifted energies and the other way around at the negative bias side (compare the closeups of the antisymmetrized spectra in Fig. 5.6(c) and Fig. 5.6(d)). Thus, it appears reasonable to assume that these features are caused by the "transversal" and "longitudinal" phonons of niobium inelastically excited by the tunneling electrons. While the dip-peak pair at about $24\,\mathrm{mV}$ is fairly clear in this spectrum, it has to be admitted that the one at about $15\,\mathrm{mV}$ appears rather as a shoulder than as a maximum. The reason for this is probably the presence of features in the low voltage range between $-10\,\mathrm{mV}$ and $+10\,\mathrm{mV}$ that are even stronger in intensity. These features produce overshoots beyond $\pm 10\,\mathrm{mV}$ affecting the signal of the "transversal" phonons expected at $\pm 15.5\,\mathrm{mV}$ As mentioned before, the magnetic field was set to be much higher than B_{c2}. It is therefore unlikely that the low bias signature is due to remains of superconductivity. Since the intensity is rather comparable to the one of the phonon excitations, it could by caused by other excitation processes of unknown nature.

In STM-IETS experiments, for example on single molecules, it is code of practice to relate the spectrum on the molecule to the one on the bare substrate in

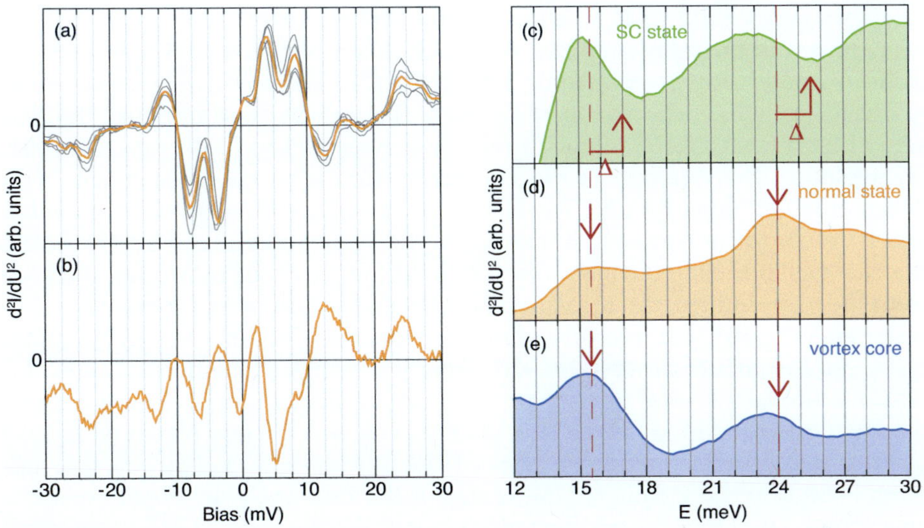

Figure 5.6: (a) Inelastic tunneling spectra in the normal state (0.8 K, 1.6 T, U_{mod}=1.8 mV) (gray) and the average of them (orange). (b) ITS spectrum again in the normal state under different tip condition (2.5 K, 1.0 T, U_{mod}=2 mV). (c)-(e) Comparison of antisymmetrisized d^2I/dU^2 spectra obtained in the superconducting state (from Fig. 5.5) and the normal state (from (a)) as well as in the Shubnikov phase in the normal conducting center of a vortex (2.5 K, 250 mT, U_{mod}=2 mV).

order to exclude that the observed features are caused by processes in the tip. In this study on the flat homogeneous surface of the Nb crystal, however, such a reference is not available. The only possibility is to compare measurements with different tip conditions. For this reason a spectrum recorded after another preparation cycle of the sample and after tip treatment is shown in Fig. 5.6(b). While the dip-peak pair of "longitudinal" phonons is still present at ± 24 mV the one of the transversal phonons, though clearly visible seems to have shifted to slightly lower energies. As discussed before this could be due to the influence of the features in the low bias range which look totally different than in Fig. 5.6(a). Regarding these difficulties, precise quantitative conclusions on the peak positions and especially on the relative intensities unfortunately remain problematic.

Tunneling spectroscopy in the Shubnikov phase

After establishing that the niobium phonons can be detected by ITS in the normal state, at least on a qualitative level, it should be checked in the following whether the same is true in the normal conducting center of a vortex. Therefore a magnetic field of $B = 250\,\mathrm{mT}$ lying in between the two critical fields was applied in the following. The magnetic flux which now enters the sample in the shape of flux lines forming a hexagonal Abrikosov lattice is presented in Fig. 5.7(a), which shows the 2f signal of the lock-in amplifier with the bias set to 1.2 mV resulting in a large difference between normal and superconducting regions. The distance of the vortices is $102\,\mathrm{nm}$ in good agreement with $98\,\mathrm{nm}$ obtained from Eqn. 5.1 for $B = 250\,\mathrm{mT}$. Fig. 5.7(b) shows 1f tunneling spectra recorded at certain distances from the vortex core in order to illustrate the changes in the superconducting DOS. At the points of maximum distance from the vortex cores, the measured spectra still exhibit a gap and quasiparticle peaks with reduced intensity upon approaching a vortex core. In the center one would expect to observe the flat density of states of the normal state. Instead an enhanced local density of states at the Fermi level is found for this distinct vortex examined here. This kind of peak was already observed in 1989 by Hess *et al.* in a STM study on $NbSe_2$ [106]. It was explained by quasiparticle bound states in the vicinity of the center of the flux line [94]. Furthermore, it was shown [107] that the conductance peak at the Fermi level is a property of superconductors in the clean limit, which is characterized by the ratio of the mean free path and the Ginzburg-Landau coherence length $l/\xi_{\mathrm{GL}} \gg 1$. In contrast, for superconductors in the dirty limit ($l/\xi_{\mathrm{GL}} \ll 1$), the enhancement at E_F was not observed and the local density of states corresponded to the one in the

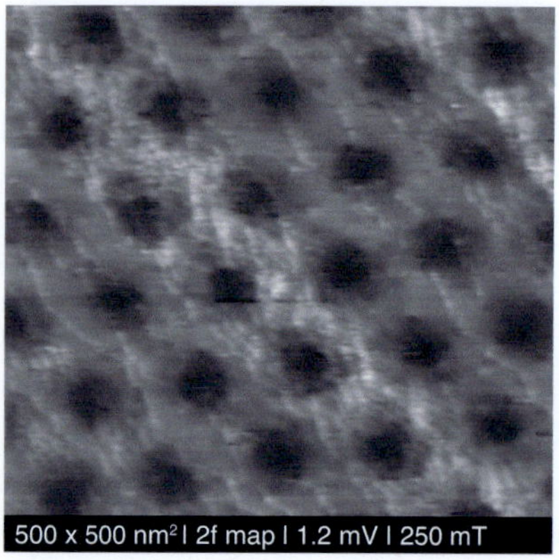

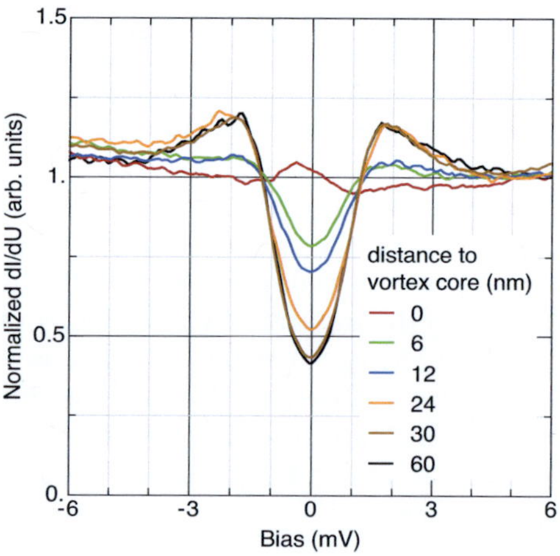

Figure 5.7: (a) Magnetic flux entering the sample in the shape of flux lines arranged in a hexagonal Abrikosov lattice. (b) Tunneling spectroscopy at the given distances to a vortex core revealing a bound state in this distinct vortex (T=0.72 K, B=0.25 T, U_{mod}=0.2 mV).

normal state. Actually, the observation of vortices exhibiting the bound state feature was rather the exception than the rule for the the Nb(110) surface under investigation here, indicating that the cleaning procedure indeed could be still improved. On the other hand, an altered density of states due to the presence of the bound state feature would, of course, greatly affect IETS.

Fig. 5.6(e), finally presents the result of IETS in a vortex core without a zero bias peak. The spectrum shows two peaks at roughly the expected energies and is thus, aside from differences in the intensity, essentially the same as the one obtained in the normal state. This kind of spectrum could be obtained only rarely. Here the fact that the surface is not the one of a bare Nb crystal turned out to be rather problematic. Surface contaminations (superstructure, NbO chains) lead to a quite strong dependence of the inelastic tunneling spectra, especially in the low voltage range, on the position of the measurement but since one has to precisely localize the normal conducting center of a vortex, which is rather difficult, the possibility of averaging spectra from different positions is hampered. Most likely it is also due to this reason that attempts of observing the transition from dips at $E_\omega + \Delta$, the phonon signatures in the superconducting state to peaks at E_ω in the normal state when measuring $\mathrm{d}^2I/\mathrm{d}U^2$ spectra along a line towards the center of a vortex did not yield unambiguous results.

In summary, one can say that the experiments on the Nb(110) surface demonstrate that the observation of the phonon modes of niobium by the spectroscopic modes of the STM is clearly feasible. The phonon signature in the superconducting density of states, which arises from strong coupling effects, can be unequivocally detected with relatively small effort. In contrast, the inelastic tunnling spectra in the normal state are much harder to obtain. One may speculate whether this is because the quasiparticles in superconducting tunneling sample a region the size of which is given by the coherence length while inelastic phonon excitations is only a very local probe [64], much more sensitive to spatially varying impurities. Nevertheless, there are measurements clearly showing the expected features albeit more statistics would help to clarify the influence of the tip condition as well as of the detailed surface structure. Last but not least, niobium is only a weak to intermediate coupling superconductor. This, of course, also implies that the possibility of phonon excitation by tunneling electrons in an IETS experiment is reduced in comparison to strong coupling materials, e.g. lead. Experiments on this material will be discussed in the following chapter.

6 Lead

In the previous chapter it was shown that STM-IETS is able to detect the phonons in niobium in the sense that clear peaks on the positive bias side of the $\mathrm{d}^2I/\mathrm{d}U^2$ *spectra appeared at the corresponding energies of van Hove singularities in this material. The normal state was achieved by applying a magnetic field larger than, at least,* B_{c1}*. Following these successful measurements, this section now presents the study of the electron-phonon coupling in lead (Pb). As an alternative approach to suppress superconductivity, the (inverse) proximity effect is introduced which occurs in thin superconductor films on top of a normal metal. Molecular beam epitaxy of lead on different substrates was employed in order to achieve a sufficiently small lead thickness. Due to the reduced dimensionality of the lead films, quantum size effects start to occur and to influence their physical properties. The results of IETS demonstrate that the full shape of the Eliashberg function of lead can be obtained by STM-IETS. In addition, exploiting the unique capabilities of the STM, the electron-phonon coupling was also probed with spatial resolution on lead areas with different local thickness revealing a distinct influence of quantum size effects. This experimental finding is supplemented by theoretical calculations.*

6.1 Motivation

Among the elemental superconductors, following niobium, lead has the second highest transition temperature being $T_C = 7.2\,\mathrm{K}$ [108]. More interestingly, with $2\Delta_0/k_BT_C = 4.3$ the ratio of the gap value and the critical temperature by far exceeds the BCS value of 3.53 and thus lead belongs to the group of so-called strong-coupling superconductors. Indeed, besides mercury, it is the element which possesses the strongest electron-phonon coupling [59]. Based on this fact, it is not surprising that deviations from BCS theory were first discovered in the measurement of the superconducting density of states of lead [23] and that McMillan and Rowell were able to extract the Eliashberg function of a real material for the first time for lead [9] as mentioned already in Chap. 3. Their result for $\alpha^2F(\omega)$ was given in Fig. 3.6. For the sake of completeness Fig. 6.1 additionally shows the dispersion relation of phonons in lead and the corresponding density of states [109]. Since lead is a very soft metal the phonon energies are quite small, e. g. compared to those of niobium, but the energy resolution of the JT-STM is still high enough to resolve the van Hove singularities at around 4 meV and 8.5 meV. The strong electron-phonon interaction, increas-

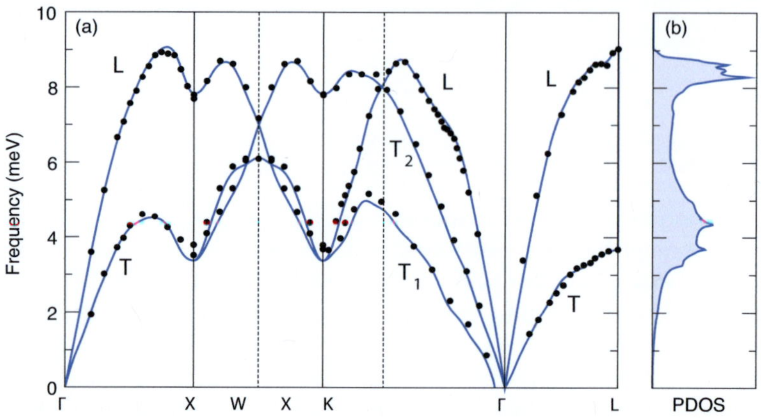

Figure 6.1: Phonons in lead: (a) Dispersion relation. Experimental data from neutron scattering (dots) and a calculation taking into account spin-orbit coupling effects (blue line). (b) Corresponding phonon density of states (PDOS) with van Hove singularities at around 4 meV and 8.5 meV. Adapted from [109].

ing the chance of excitation of phonons also in the normal state, is the motivation to investigate lead in this study as well. Furthermore, the large number of existing experimental and theoretical results allows for a comparison.

All the experiments on lead in this study were carried out not on a bulk specimen but on thin films of the material deposited by MBE onto copper or silicon substrates. The reason for considering thin films is twofold. First, by shrinking the thickness of the system in the direction perpendicular to the sample surface, the reduced dimensionality should give rise to quantum size effects in the electronic density of states as well as in the density of states of the phonons. While the former can be understood in terms of quantum well states and is very well known from many studies during the last decade (as is presented in more detail in the next section), the latter would even offer the possibility of determining the dispersion relation of lattice vibrations by STM. The idea of such a measurement is based on the fact that in a thin film of certain thickness only excitations with the corresponding wavelengths are allowed and, thus, instead of the continuous density of states one obtains discrete peaks which can be matched to individual momenta of the excitations. The dispersion relation of magnons in antiferromagnetic manganese films was successfully determined in this manner by Gao *et al.* [53].

Another reason to examine thin films is that also the superconducting properties may vary drastically upon reducing the sample size. For example, in agreement with Ginsburg-Landau theory bulk lead is a type-I superconductor since $\kappa = \xi_{GL}/\lambda_L \approx 39\,\mathrm{nm}/87\,\mathrm{nm} = 0.45 < 1/\sqrt{2}$. The critical field is

$B_C \approx 80\,\mathrm{mT}$ [108]. In thin films the Ginsburg-Landau coherence length may be limited by the film thickness, thus driving $\kappa \geq 1/\sqrt{2}$ changing the superconducting behavior from type-I to type-II which was actually observed in lead films deposited on a Si(111) surface [110–113].

The fact that superconductivity persists at all for films much thinner than the bulk coherence length [111, 114, 115] is yet more interesting because it contradicts the Mermin-Wagner theorem [7, 116–118]. Experimentally, however, it was found that even a film of the ultimate thickness of only one monolayer remains superconducting [119]. Before it was speculated that quantum fluctuations may prevent the thin film from becoming superconducting. The detailed influence of the reduced thickness of lead on important superconducting properties like gap size or transition temperature, however, is still a controversial issue. While some experiments report a decrease of T_C upon reducing the film thickness [60, 110, 120], other studies found no certain trend but an oscillation in T_C which was ascribed to quantum size effects since it correlates with the variation of the electronic density of states induced by the already mentioned quantum well states [114, 115].

The experiments discussed so far all were done on lead films fabricated on substrates made of silicon which, as a semiconductor, provides no density of states at the Fermi level and consequently the electrons are trapped in the superconducting film. The latter remains also true for highly doped silicon after having passed through its metal-insulator transition [121] which has to be employed in all cases of STM studies at low temperatures.

This situation changes when a normal metal substrate is used and one has to account for the proximity effect which occurs at the interface between a superconductor and a normal conducting material. Neglecting the proximity effect, n_S, the density of Cooper pairs in the superconductor, has to go down to zero at the interface with the normal metal. This decay takes place on a length scale given by the Ginsburg-Landau coherence length ξ_{GL} specifying the minimum distance on which the superconducting order parameter Ψ and thus also n_S can change. At a real interface, however, Cooper pairs are able to diffuse into the normal metal and hence a finite order parameter is induced there, which decays on another characteristic length scale termed extrapolation length [7, 122]. While proximity effect usually expresses this induction of superconductivity in the normal metal, it also works the opposite way as reduction of the superconducting transition temperature which depends on the ratio of the thickness of the normal metal and of the superconducting layer [123]. Experimentally it was found that T_C rapidly goes to 0 when the the thickness of a lead film on an extended copper substrate is less than $d_{Pb} \approx 350\,\mathrm{nm}$ [124].

Further, Hilsch showed that this critical thickness is smaller when the residual resistance of the normal metal substrate is higher, e. g. $d_{Pb} \approx 8\,\mathrm{nm}$ for a cold-condensed copper substrate [124]. This explains why the lead films on highly doped Si:As are still superconducting. The substrate is indeed metallic but the electronic density of states at the Fermi level crucial for the proximity effect is still by far smaller than in copper.

The different superconducting behavior of thin lead films on copper and on silicon substrates allows to compare two different approaches to achieve the normal state necessary for the study of electron-phonon coupling by STM-IETS. First, a magnetic field is employed to suppress the superconductivity of lead on silicon. In the alternative system, lead on copper, this step is dispensable since superconductivity is already suppressed by the inverse proximity effect. An additional experiment, where one atomic layer of copper nitride is introduced between the copper substrate and the lead film addresses the question whether this insulating layer is appropriate to prevent the inverse proximity effect. However, before presenting the experiments and their results the quantum size effects that appear in thin lead films on all substrates have to be considered since they have a major influence on many properties, especially the growth.

6.2 Quantum Well States

The Fermi wave length $\lambda_F = h/\sqrt{2mE_F}$ of a common metal is several angstrom and thus of the same order of magnitude as the interatomic distance in the material [7]. Thus the energy levels of electrons confined in a film of a thickness of only a few monolayers (ML) are quantized. In the present case of lead films deposited on a substrate the confinement in the direction perpendicular to the sample surface on the vacuum side is naturally provided by the work function (4.25 eV). Whether the substrate leads to a confinement on the other side depends on the availability of state there for electrons traveling along the normal direction. This is illustrated in Fig. 6.2(a) showing the dispersion relations in [111] direction of both materials for the example of lead on Cu(111) [125]. In copper there are no states for energies $-1.0\,\text{eV} < E < 3.8\,\text{eV}$ in this direction and thus the *sp*-electrons of lead cannot propagate into the substrate [125].

The simplest model to describe an electron confined in the lead film is as a *particle in the box*. In this model one considers a potential well of infinite depth with a width given by the thickness of the lead film $d_l = ld_{\text{ML}}$ where l is the number of atomic layers and d_{ML} the thickness of one monolayer. For lead in the [111] direction $d_{\text{ML}} = 2.86\,\text{Å}$ [125]. Barriers of infinite height only allow wave functions that are exactly zero at both boundaries which is fulfilled only if multiples of half the wavelength match the width and so the quantization condition for the z-component of the wave vector reads

$$k_{ln} = \frac{n\pi}{ld_{\text{ML}}}, \qquad n \in \mathbb{N}. \tag{6.1}$$

While the electron motion in the film plane is not affected by this quantization it selects from the continuum of states in the [111] direction only those which have appropriate wave vectors as indicated in Fig. 6.2(a) for the example of 16 ML. The energy of the resulting quantum well states (QWS) can be calculated from

$$E_{ln} = \frac{\hbar^2}{2m}\left(k_{ln}^2 - k_F^2\right) \tag{6.2}$$

by assuming the quadratic dispersion of quasi-free electrons where $k_F \approx 1.59$ Å^{-1} is the absolute value of the Fermi wave vector known from bulk lead [126–128]. Fig. 6.2(b) shows the energy of the QWS calculated from Eq. 6.2 for 1 to 24 atomic layers. The almost vertical hyperbolas give the energy as a function of continuous film thickness for discrete quantum numbers n while the open dots mark the points where the thickness is an integer multiple l of the lead layer spacing in [111] direction. Whenever the ratio n/l takes the same value, the corresponding QWS have the same energy. This is the reason why there is a state at a constant energy of $0.71 \pm 0.06\,\mathrm{eV}$ for all films comprising an even number of atomic layers marked by the horizontal blue line. The other QWS

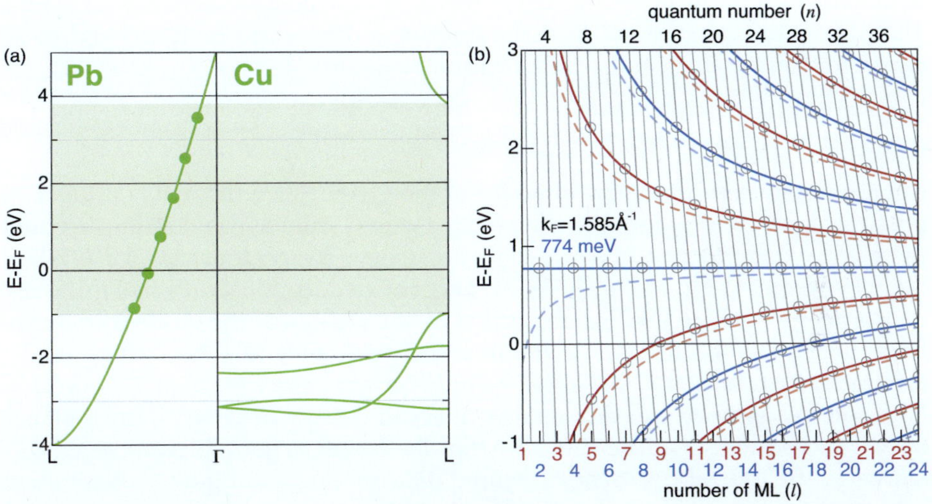

Figure 6.2: Quantum well states in thin Pb films: (a) Sketch of the dispersion relation for Pb on Cu in the [111] direction. Electrons cannot pass from Pb to Cu for energies between $-1.0\,\mathrm{eV} < E < 3.8\,\mathrm{eV}$ due to the gap in the projection of the copper bulk bands in [111] direction. The resulting discretization is indicated by the circles on the bulk band for the example of 16 Pb layers. Adapted from [125]. (b) Calculated QWS (open circles) in dependence of the Pb layer thickness l and the quantum number n. Blue and red lines are guides to the eye indicating the evolution of QWS with increasing thickness for even and odd layers, respectively. The dashed lines illustrate the effect of an additional phase at the example of a mean surface shift of $\delta_0 = 0.05\, d_{\mathrm{ML}}$.

are located on increasing or decreasing hyperbolas for energies below or above the energy of the constant state, respectively, so that the separation of the states gets smaller for thicker films. As a guide to the eye indicating the evolution of a certain QWS with increasing thickness, these hyperbolas are also colored in blue for even l and in red for odd l.

Interesting physical consequences are expected to occur when a QWS lies close to the Fermi level. The first time a QWS passes through E_F is at 9 ML but also for larger thicknesses QWS keep crossing the Fermi level at certain numbers of layers. One might argue that the assumptions made by the infinite square well model are to crude in order to describe the real system correctly. It is certainly true that the depth of the potential well being about 4 eV is far from infinity and thus the model is not able to treat the electron spill out into the vacuum and the copper region. On the other hand it was found experimentally that this simple model matches the measured energies very well, at least in an interval of ± 1 eV around the Fermi level. In particular, Otero *et al.* also observe a state at a constant energy of about 0.65 eV for every second layer as predicted by the infinite square well model [125]. A more realistic theoretical picture can by obtained by a phase accumulation model for the confinement barriers [129]. The phase shifts accumulated at the limiting surfaces can be transfered to an altered effective width and with a mean surface shift δ_0

$$k_{ln} = \frac{n\pi}{l d_{\mathrm{ML}} + \delta_0} \tag{6.3}$$

can accurately reproduce the energies of the QWS. Since the real confinement potential is soft and not square-shaped one would still have to take into account the energy dependence of δ_0 in order to achieve agreement also for QWS at higher energies [129]. In Fig. 6.2(b) the effect of an additional effective thickness on the energies of the QWS is illustrated by the change of the guide-to-the-eye-hyperbolas (dashed lines). While, in principle, they could be shifted almost arbitrarily by tuning the parameter δ_0, only certain values ($\delta_0 = m\,2/3\,d_{\mathrm{ML}}, m \in \mathbb{N}$) result in a state of constant energy at about 0.7 eV. In general, the distance of two subsequent QWS becomes smaller the larger δ_0 gets. In some sense this allows to "fit" the model to the measured data but an unambiguous assignment is hard to achieve. This makes it difficult, in particular, to determine the exact height of a film exclusively from a measurement of the QWS and therefore, if possible, also information from the topography should be considered.

Experimentally, QWS can be accessed *e. g.* by angle-resolved photoemission (ARPES) [130, 131], He atom scattering (HAS) [132], but also in real space by STM. Measurements of $\mathrm{d}I/\mathrm{d}U$ spectra were reported for lead films of different thicknesses on a Cu(111) substrate [125] as well as on Si(111) [133]. Furthermore, it was demonstrated that the measurement of $\mathrm{d}z/\mathrm{d}U$ reveals spectra that show peaks at almost the same voltages corresponding to QWS energies as the differential conductance [134]. z-U curves are obtained by leaving the feedback loop active while ramping the voltage. The advantage is that the current does not rise exponentially when the voltage approaches the work function, thus allowing to record QWS over the whole energy range of interest. A drawback is that the tip might contact the sample at $U = 0\,\mathrm{V}$.

The studies cited above discovered a distinct influence of the QWS on the growth of lead films which is basically a double-layer mode, i. e. in a certain thickness range even layers are preferred over odd ones or vice versa. The preferred numbers of layers were therefore named "magic heights" [135, 136]. This special growth behavior can be understood from the positions of the QWS. Whenever the highest occupied QWS comes close to E_F the total energy is enhanced compared to the one of the neighboring layers which then, in general, provide QWS much further below the Fermi level as is obvious from Fig. 6.2. This leads to the instability of the corresponding number of layers and thus, when growing, the system avoids this thickness and bifurcates into the neighboring more stable layers. In particular, a film becomes most unstable if a QWS is right at E_F which for example is the case for 9 ML and appears again with a periodicity of about 9 ML. At these layers, moreover, there is a change of the film stability from even to odd or the other way around [126]. For larger thicknesses (more than 20 ML), however, the quantum effects on the growth become less and less significant. This growth effects characteristic could be confirmed in this study as presented in the following section. In addition to their influence on the growth, QWS also affect other physical properties like electrical resistivity or the work function as was already predicted in 1975 by Schulte [137], who was the first to consider QWS in thin metallic films theoretically. In the meantime many of these quantum size effects have been confirmed experimentally. Most interesting with respect to this thesis is the observation of an oscillating superconducting transition temperature as well as an oscillatory electron-phonon coupling [114, 120, 126] in lead films on Si(111).

6.3 Sample Preparation and Characterization

6.3.1 Preparation of Silicon, Copper and Copper-Nitride Substrates for Lead Deposition

Before the deposition of lead, a suited surface condition of the respective substrate has to be achieved. The relevant steps for the different substrates used in the study are now described.

Silicon

Lead on silicon is a prototype of a nonreactive, epitaxial metal-semiconductor contact [138] and has been widely studied in the past. However, as already mentioned in Sec. 6.1 , the doping level of silicon singe crystals has to be high enough so that the metal-insulator transition is passed because a finite conductance of the sample is of cause mandatory for STM experiments at low temperatures (even though only few relevant publications mention this point [111]). The crystals for the present study were rectangles of $4 \times 9\,\mathrm{mm}^2$ scratched by a diamond and afterwards broken from a n-type Si(111) waver doped with phosphorus and possessing a resistivity of 1 to $5\,\mathrm{m\Omega cm}$.[1] Each crystal was fixed indirectly to a molybdenum sample holder by spot-welding thin tantalum stripes as belts at the left and right side of the sample. During the preparation silicon should be only handled with plastic tools since the contact with metals might cause undesired surface reconstructions.

At ambient conditions the surface of silicon is contaminated by oxide and carbide which has to be removed inside the UHV chamber. Ion sputtering, however, produces undesirable radiation damage at the surface and point defects often remain even after annealing [139]. That is why high temperature thermal etching at about 1200 °C is usually applied in order to get a clean surface [140]. The sample has to be carefully degassed at 780 °C for several hours and than flashed to the end temperature of 1200 °C for a maximum time of 1 min in order to get rid of the oxide on the one hand but to avoid carbon diffusion from the bulk and the formation of silicon carbide on the surface on the other hand [138, 141]. Since the development of a uniform surface is expected to take some time the temperature is only decreased to 810 °C and kept there for some minutes before the sample is allowed to slowly cool down to room temperature. In the used setup the heating was achieved by electron bombardment from the backside of the sample holder where a hole allows direct impact of the electrons on the silicon. Since the melting temperature of silicon is only 1414 °C [142] the sample might slightly melt at the backside which had no negative effect on the subsequent experiment.

Even though the high temperature used in this method is known to bear the risk of changing the doping profile [138, 139], it worked well in the present case as shown in Fig. 6.3(a) which presents a LEED pattern of the famous 7×7 reconstructed Si(111) surface. This arrangement of surface atoms is an equilibrium phase [79, 143] that forms on a clean Si surface in a temperature range between 200 to 500 °C [144, 145] and 860 °C [146] which is naturally reached in the present preparation procedure when cooling down from 1200 °C. The good quality resulting from the preparation is further verified by the STM image shown in Fig. 6.3(b). The appearing vacancies might originate from the high concentration of dopants.

A cumbersome issue was found concerning the the re-use of silicon crystals after lead was once deposited. Even though the preparation procedure de-

[1] Attempts using n-type silicon with a resistivity of 1 to $10\,\Omega\mathrm{cm}$ failed since it was found that those crystals are not conducting at temperature below a certain temperature estimated to be around 14 K in this case.

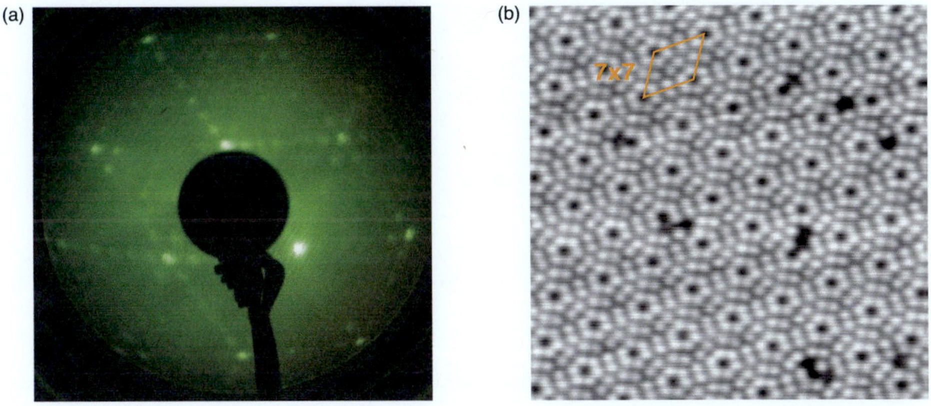

Figure 6.3: Quality of the Si(111)-(7×7) surface: (a) LEED pattern. (b) STM image (17.6×17.6 nm^2, 100 pA, 1 V, 23.2 K).

scribed above was obviously able to remove lead from the surface as could be confirmed by LEED, the presence of lead during heating must have a distinct influence on the doping level since the crystals were found not to be conductive anymore at low temperatures. Therefore it was necessary to prepare completely new silicon crystals for each deposition of lead.

Copper and copper-nitride

As a alternative metallic substrate, on which a thin lead film is expected to be in its normal state due to the proximity effect, Copper was chosen. Copper exhibits a face-centered cubic crystal structure with a lattice constant of 361 pm [147]. The interlayer distance in the $\langle 100 \rangle$ direction is thus half this value and in the $\langle 111 \rangle$ direction it is 208 pm. Since the solubility of the elements lead and copper is negligible [148], there is no alloying of lead and the substrate atoms at the interface which is important concerning the purity of films after deposition. The choice of copper as a substrate has also a practical aspect. The [111] as well as the [100] surface of a copper single crystal can be easily cleaned by Ar^+-sputtering at an energy of 1.5 keV for about 30 min followed by an annealing step in which the crystal is heated to 450 °C with a temperature raise of about 1 °C/min. Even if the crystal was exposed to air or a huge amount of lead was deposited before, repeating this procedure only a few times is sufficient in order to achieve reasonable large terraces free of contaminations.

The experiments on lead on copper should be further supplemented by a study of lead deposited on a film of copper-nitride on Cu(100). The main reason for the modified substrate was the possibility that the lead film decoupled from

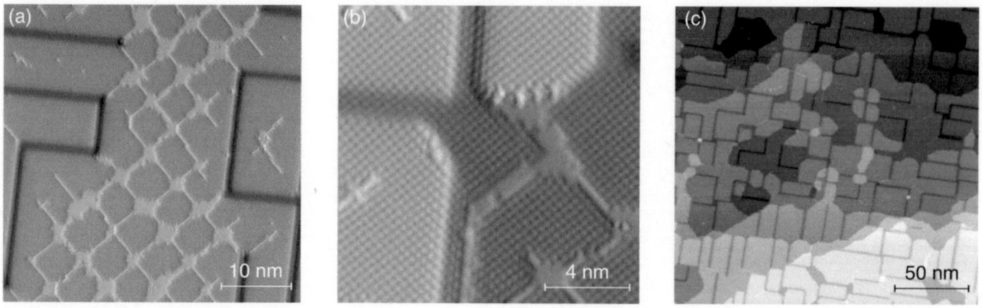

Figure 6.4: STM images of CuN on Cu(100): (a) At intermediate coverage square-shaped islands with bare copper in between coexist with totally covered areas where broader tranches release the strain due to lattice mismatch. As can be seen from the atomically resolved closeup in (b), the broad tranches are also covered by CuN. (c) Large scale STM image of a sample at high coverage. The whole surface is covered by CuN.

the metallic substrate by the insulating buffer layer could still be superconducting. The decoupling ability of CuN/Cu(100) was already demonstrated for the magnetic properties of single atoms [149]. Those atoms lay on small CuN islands of about $5 \times 5\,\mathrm{nm}^2$ which are arranged in a square pattern. While this arrangement is well suited for decoupling single atoms from the metallic substrate it cannot be used for decoupling an extended film because there are metallic regions in between the islands. For the purpose of the present study, a closed film is thus the goal of the preparation of CuN/Cu(100).

After the crystal had cooled down to room temperature after the cleaning process, the preparation of CuN is performed by exposing the clean Cu(100) surface for 15 to $20\,\mathrm{min}$ to a nitrogen ion beam of $500\,\mathrm{eV}$ produced by the sputter gun when setting up a partial pressure of nitrogen in the lower $10^{-4}\,\mathrm{mbar}$ range in the preparation chamber [150]. This preparation step again coarsens the surface and thus the sample has to be annealed a second time, though one has to consider the decomposition temperature of copper-nitride as a metastable material being 250 to $300\,°\mathrm{C}$ [151]. The temperature of $600\,\mathrm{K}$ given in [150] thus turned out to be slightly too high and an annealing temperature of 575 to $585\,\mathrm{K}$ was found to be adequate for the present setup.

Fig. 6.4 shows STM images of CuN on Cu(100) for different amounts of nitrogen. In agreement with previous studies [150, 152], for low coverages the topography is characterized by many more or less square shaped islands. These islands are oriented along the $\langle 100 \rangle$ directions of the Cu(100) surface and make up a regular pattern with stripes of bare copper in between. In the STM image these stripes appear higher than the islands because of a much higher density of states. When the coverage approaches a complete layer the islands start to coalesce and build up larger joint areas of CuN. These areas are interrupted by trenches, 1 to 2 nm wide, running along the ⟨110⟩ directions which release the strain in the surface resulting from the lattice mismatch of CuN and copper [150, 152]. Fig. 6.4(a) shows a STM image of a sample with intermediate coverage where both topographies coexist. Most importantly, as one can see from the STM image in Fig. 6.4(b) the tranches are also completely covered by CuN so that a topography as it is shown in Fig. 6.4(c) provides a sample surface where the metal is totally concealed by the insulating layer. The terraces of the the CuN/Cu(100) surface are flat over wide areas though there are some small islands or depressions of a height corresponding to the Cu(100) atomic layer distance which often occur at ends or interception of the trenches. In general, it seems that the growth of a lead adlayer might be affected by this suboptimal substrate topography. Further it turned out, that the detailed adjustment of the parameters (time, pressure, annealing temperature) leading to a full coverage is difficult to reproduce and thus, in the beginning, the CuN surface was checked by STM each time before lead was deposited.

Sublimation and deposition of lead

Lead atoms are sublimed from a lead wire of high purity (99.9985 %, Alfa Aesar) mounted in the MBE evaporation source described in Chap. 4. Due to the low melting point of lead (327 °C [153]) the lead wire has to be placed in a crucible made of molybdenum and the whole evaporation device has to be mounted in an upright orientation.

The respective total amount of deposited lead was controlled by the deposition time after the specific deposition rate had been calibrated from STM images for a certain current to the flux electrode. In all experiments the latter was regulated to 600 nA, which was achieved by an emission of 7.5 mA with the high voltage set to about 280 V, resulting in a deposition rate of about 2 ML/min.

In this study, deposition of lead should take place always at room temperature. Therefore it was necessary to wait until the sample had cooled down after the cleaning procedure. After the deposition of lead the samples were immediately transfered to the STM.

6.3.2 Characterization of Lead on Different Substrates

Lead on Si(111)-(7×7)

In agreement with previous studies [60, 114, 126, 154], where lead was deposited on the Si(111)-(7×7) surface at room temperature, flat top wedge-like islands were observed (Fig. 6.5(a)). The growth mode is of the Stranski-Krastanow type, i. e. the three-dimensional islands appear on top of a wetting layer (WL).[2]

[2] A layer-by-layer growth can be compelled by cooling the substrate by liquid N_2 to temperatures below 150 K during Pb deposition and subsequent annealing to room temperature [126].

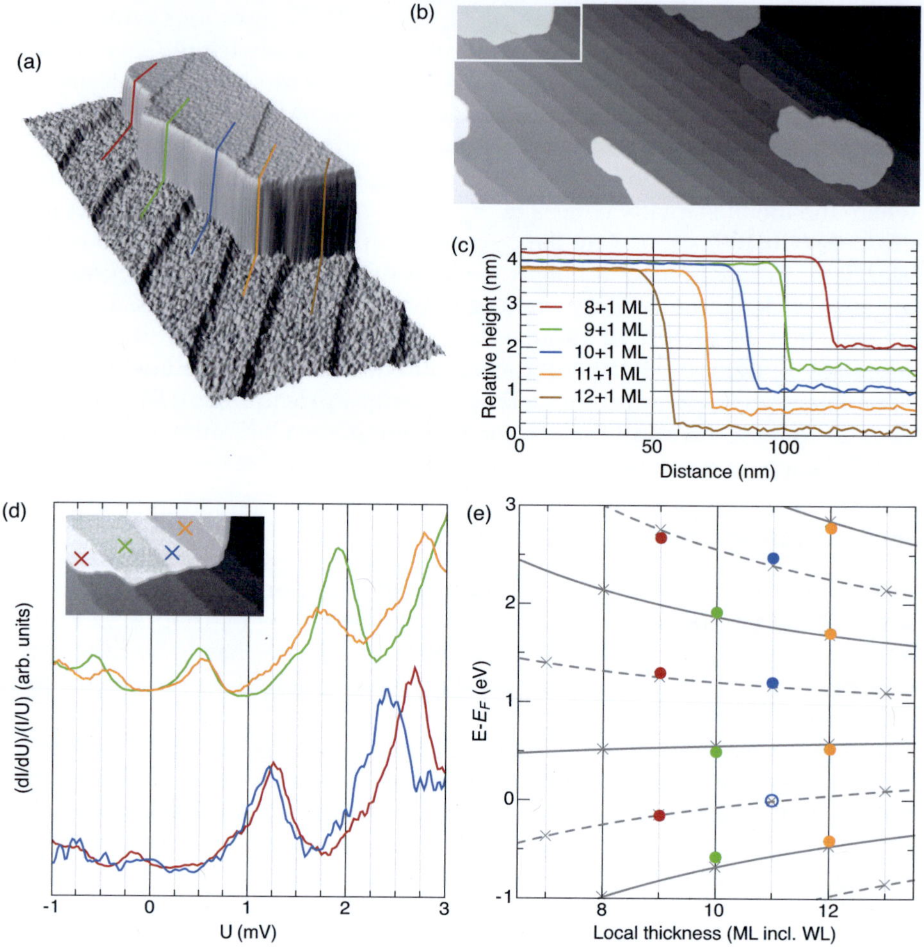

Figure 6.5: Topography of Pb islands on Si(111)-(7×7): (a) 3D view of a Pb island taken from the upper left of the large-scale STM image (1500 × 650 nm^2, 100 mV, 100 pA) presented in (b). The island grows as a flat-topped wedge on a disordered Pb wetting layer which forms on top of the stepped Si(111)-(7×7). (c) Height profiles along the lines indicated in (a). The local thickness with respect to the wetting layer corresponds roughly to multiples of the Pb interlayer distance in ⟨111⟩ direction given in the legend. The minor height variation on top of the island is due to the mismatch of the lattice constants of Si and Pb as well as due to different electronic density of states caused by QWS. (d) Measurement of QWS via STS at positions of different local thickness as indicated in the inset. (e) Comparison of the peak positions from (d) to the infinite square well model. An additional phase corresponding to a mean surface shift of δ_0=0.75 ML was assumed in order to improve the agreement. The solid and dashed lines indicate the evolution of QWS for even and odd layer, respectively.

The wetting layer formed on the Si(111)-(7×7) surface is disordered and appears quite rough.[3]

The islands are Pb single crystals with their ⟨111⟩ axis perpendicular to the substrate [127, 138]. The large scale STM image in Fig. 6.5(b) shows that the Pb islands often grow along the Si step edges and extend in their direction to some 100 nm but many of them are also wide enough to span several of the 50 to 200 nm wide terraces. At the Si steps the local thickness of the island changes by one Pb layer resulting in the observed wedges. For an STM study the formation of 3D islands is convenient since the local thickness referenced to the wetting layer can be determined from the height profile of an island as presented in Fig. 6.5(c) for the lines indicated in (a). The different parts of the Pb island, which spans five Si terraces, have relative heights with respect to the corresponding underlying terraces that are roughly multiples of $d_{Pb111} = 286\,\mathrm{pm}$, the Pb interlayer distance in ⟨111⟩ direction. Accordingly, the local thickness of the island in Fig. 6.5(a) changes from 9 to 13 ML including the wetting layer. The minor height variations on top of the island originate from the different interlayer distance of Pb and the Si substrate ($d_{Si111} = 384\,\mathrm{pm}$[4]) as well as from different density of states which is modulated by QWS. [154].

Especially the latter effect is responsible for the alternating contrast from even to odd layers as apparent in the top view of the island in the inset of Fig. 6.5(d). The spectra presented in this graph were obtained from I-U curves recorded on the indicated locations on the island by numerical differentiation and normalization to the total conductance. The enhanced density of states at the energies of the QWS is clearly visible. As shown in Fig. 6.5(e) the peak positions almost perfectly match the infinite square well model if the thickness of the wetting layer is assumed to be 1 ML as was already concluded in a previous study [60]. Introducing an additional phase corresponding to a mean surface shift of $\delta_0 = 0.75\,\mathrm{ML}$ slightly improves the agreement.

Lead on Cu(111)

The growth of Pb on the Cu(111) surface is quite similar to the one on Si(111)-(7×7). The growth mode for the substrate at room temperature is again of the Stranski-Krastanov type [155] and 3D islands start to appear once a wetting layer is completed. The wetting layer, however, is more smooth in this case [156].

Concerning the islands, also wedge-like structures were observed when spanning several Cu terraces. The wedge presented in Fig. 6.6(a) for example consists of 14 ML (including the wetting layer) on the lowest terrace and the height

[3] An ordered interface can be obtained by forming the Si(111)-($\sqrt{3}\times\sqrt{3}$)$R30°$-Pb phase [138] before growing the islands, though the latter were found to behave similar to islands on the disordered substrate [60].

[4] Si exhibits a face centered cubic crystal structure with a lattice constant of $a_{Si} = 543\,\mathrm{pm}$ [142].

is reduced by one Pb layer at every subsequent step. The number of layers was extracted from the height distribution in Fig. 6.6(b). The step heights on the wetting layer are 2.1 Å and thus match the interlayer distance of Cu in ⟨111⟩ direction. The heights of the different Pb areas with respect to the corresponding underlaying terraces roughly agree with multiples of the Pb layer separation $d_{\mathrm{Pb111}} = 286\,\mathrm{pm}$. The deviations can be again explained as an effect of the

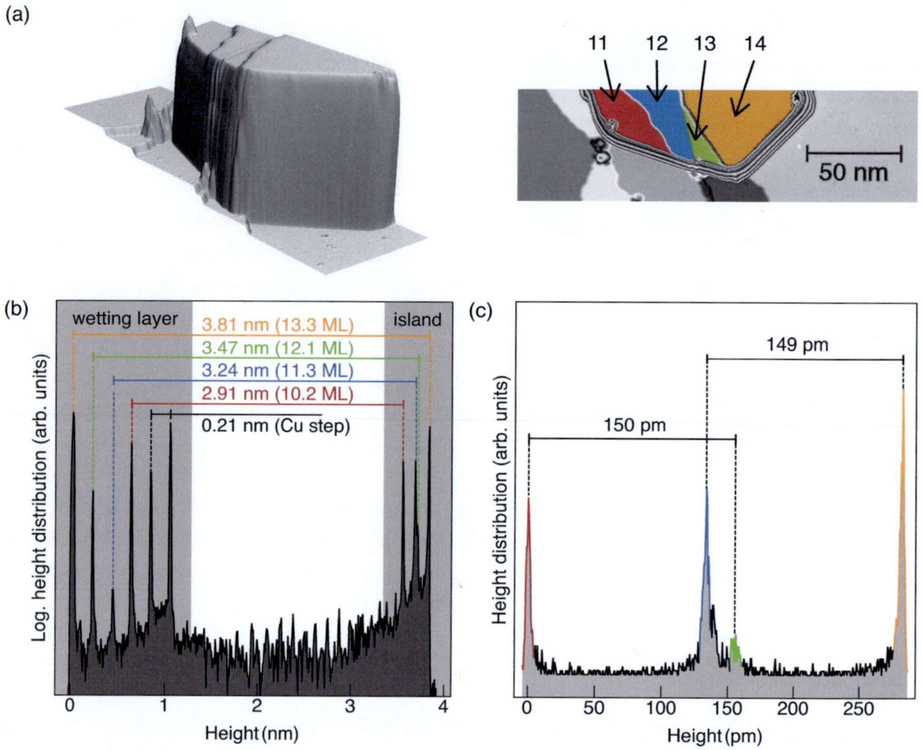

Figure 6.6: Pb wedge on the Cu(111) surface: (a) 3D and top view of a wedge-like Pb island on the stepped Cu(111) surface (1 V, 100 pA). Areas of different local thickness are marked with different colors in the top view. (b) Height distribution of the image presented in (a). The peaks in the wetting layer area correspond to the Cu terraces separated by 2.1 Å. The relative heights of Pb areas with respect to the underlaying terraces are indicated and roughly match multiples of the Pb interlayer distance. The total height indicated in a also includes the WL. QWS are responsible for the bilayer modulation of the apparent height on top of the island due to different integrated DOS. Only the height difference between every second Pb area agrees with the expected value of $2(d_{\mathrm{Pb111}} - d_{\mathrm{Cu111}})$ as shown in (c).

electronic density of states modulated by QWS. A measurement of the latter is presented later in Fig. 6.16(c). The image in Fig. 6.6(a) was recorded at a bias of 1 V. Since the integrated density of states at this energy is larger for the even layers the tip does not have to get as close to the surface as in the case of odd layers and thus they appear higher. As a consequence, the step heights on the island are not given by $d_{\mathrm{Pb111}} - d_{\mathrm{Cu111}} = 77\,\mathrm{pm}$, as one would estimate from simple topographical considerations. Only when considering the height difference between every second layer, where the QWS are at the same or comparable energies, the expected value is found as twice the difference of the lattice constants (Fig. 6.6(c)).

In contrast to the situation of Pb on Si(111), the appearance of these wedge-like islands is the exception. The majority of islands indeed spans several Cu terraces as well but stays at the same height as can be seen in the large scale STM image of a sample with a nominal total coverage of 3.8 ML in Fig. 6.7(a) where the Pb areas are marked with different colors according to their local thickness. The distribution of the local thickness presented in Fig. 6.7(b) is in agreement with the findings of Otero *et al.* [136] and illustrates that the distribution is not statistical but certain heights are strongly preferred. Most of the islands have a constant thickness of 6 or 8 ML on top of the wetting layer, which are thus the "magic heights" for this amount of Pb. As a consequence, rather than forming flat-topped wedges it even seems more favorable for the islands to decrease their local thickness by 2 ML at an upward directed step in order to end up with a preferred thickness again. Only where there are closely spaced steps it appears that the narrow Cu terraces help to stabilize unfavorable heights. In particular, there are only two rather small regions in Fig. 6.7(a) that are 11 ML thick and 9 ML was not observed at all.

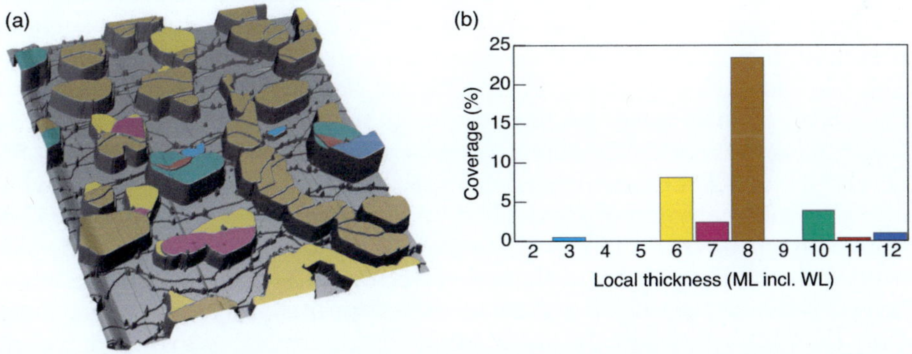

Figure 6.7: (a) STM topography (800×800 nm^2, 1 V, 100 pA) showing Pb islands on the stepped Cu(111) surface. The colored surfaces indicate the local thickness the distribution of which is given in (b).

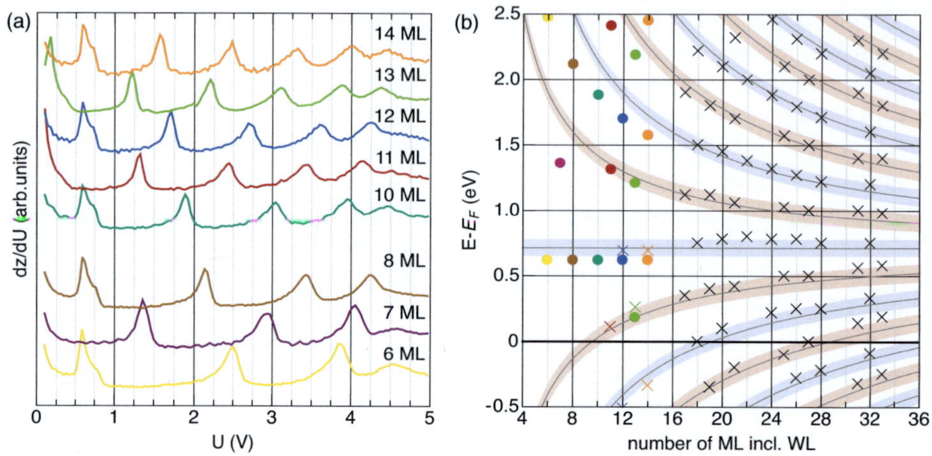

Figure 6.8: QWS in Pb on Cu(111): (a) Numerically derived z-U spectra ($I_{\mathrm{set}} = 1\,\mathrm{nA}$) revealing QWS for every local thickness observed on a sample with a nominal coverage of 3.8 ML. (b) Comparison of measured QWS to the infinite square well model. The widened blue and red guides to the eye indicate the energy interval corresponding to $k_F = 1.590 \pm 0.005\,\mathring{A}^{-1}$ for even and odd layers, respectively. Peak positions obtained from $\mathrm{d}z/\mathrm{d}U$ and from $\mathrm{d}I/\mathrm{d}U$ spectra are marked by dots and crosses, respectively. The black markers refer to a sample with a nominal Pb coverage of 16 ML.

Fig. 6.8(a) shows the measurement of QWS for every local thickness present in Fig. 6.7(a). This time the spectra were obtained as numerical derivatives of z-U curves which turned out to be the fastest and most convenient way of getting access to the peak positions. Actually, after justification with respect to the height determination from the height profiles, this kind of measurement was used to attribute the local thickness to all the different areas in Fig. 6.7(a). The QWS extracted from $\mathrm{d}z/\mathrm{d}U$ spectra are depicted by the colored dots in Fig. 6.8(b). In this figure also the maxima of the $\mathrm{d}I/\mathrm{d}U$ spectra on the wedge (Fig. 6.16(c)) are marked as colored crosses and the black crosses indicate QWS observed on thicker islands on another sample (Fig. 6.17(a) and (d)) onto which a total nominal amount of 16 ML had been deposited. The positions of QWS from both types of measurements agree well with the infinite square well model even without applying an additional correcting phase, at least for the thicker layers. However, the $\mathrm{d}z/\mathrm{d}U$ values seem to appear slightly too low in general. Only the 7 ML value deviates substantially. Furthermore, as expected, the infinite square well model describes the QWS close to E_F better than the ones at higher energies.

From this graph it becomes clear that a height of 9 ML does not appear at all since this would imply an occupied QWS just below the Fermi level, unfavor-

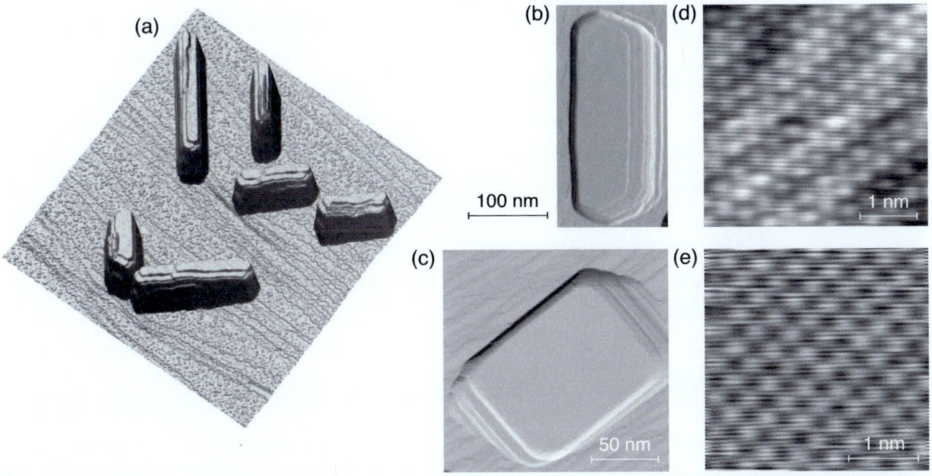

Figure 6.9: Pb islands on Cu(100): (a) Elongated islands in a large scale STM topography ($940\times940\,nm^2$, 1 V, 100 pA). (b) and (c) show a $\langle 111\rangle$ and a $\langle 100\rangle$ oriented island, respectively. The different orientations are confirmed by the atomically resolved closeups in (d) and (e).

able in terms of total energy. Still in the case of 11 ML, the close vicinity of the now unoccupied QWS to E_F leaves this height unfavorable compared to the neighboring layers. In conclusion one can say that the QWS, the observation of "magic heights" as well as their distribution are in total agreement with the previous findings on Pb on Cu(111) by Otero *et al.* [136].

Lead on Cu(100)

A different symmetry of the substrate surface has a large influence on epitaxial growth. In order to attain Pb islands of different geometry, thus, in addition a copper crystal of $\langle 100\rangle$ orientation was used. Fig. 6.9(a) shows the STM topography of Pb islands on the Cu(100) surface. A noticeable characteristic is the pronounced elongated shape of the majority of the islands. The length extends to some 100 nm while the width is only several 10 nm. Actually, in the STM images the dimensions seem to be larger but the features at the island edges originate from the convolution of the actual topography, which comprises perpendicular "walls" that are several nanometer high, with the distinct shape of the STM tip especially influencing the apparent width. In this sense only the flat top of e. g. the island in Fig. 6.9(b) reflects the real lateral shape of the object.

Again the islands grow on top op a wetting layer which, this time, possesses a well ordered $c(5\sqrt{2}\times\sqrt{2})R45°$ reconstruction. The elongated shape of the surface unit cell of the wetting layer was found to be responsible for the orientation of the Pb islands which thus only show two directions perpendicular to each other [157]. The $c(5\sqrt{2} \times \sqrt{2})R45°$ structure is only one of several interesting surface alloys and reconstructions occurring for a sub-monolayer amount of Pb on Cu(100) [158–160]. Those were all observed during the experiments but are not further discussed here.

Sometimes, though rarely, another island geometry was observed (Fig. 6.9(c)). While the elongated islands grow with their $\langle 111 \rangle$ direction perpendicular to the surface which is obvious from the $120\,°$ angles at the edges as well as from the atomically resolved surface (Fig. 6.9(d)) which shows a hexagonal pattern, the other kind of islands exhibits angles of $90\,°$ at the corners and the closeup of the surface in Fig. 6.9(e) reveals a fourfold symmetry. Hence, those islands are oriented with their $\langle 100 \rangle$ axes perpendicular to the surface. Unlike the appearance of elongated islands, which is well known from previous studies [157, 161], the (100) islands were not reported in literature so far. In principle, the appearance of islands of two different orientations would allow to study dependencies on the crystallographic direction on the same sample.

In contrast to the situation on Cu(111), Pb islands on Cu(100) are much higher. The islands in Fig. 6.9(b) and (c) exhibit heights of about $5\,\mathrm{nm}$ and 3.9 to $10\,\mathrm{nm}$, respectively. This means that the thickness is at least 17 ML but most of the observed islands rather consist of much more than $20\,\mathrm{ML}$. Hence, the observation of quantization effects is unlikely on this sample. An attempt of reducing the island height by depositing less Pb was not successful, only the amount of islands decreased. The only way of achieving thinner islands would probably be a reduced substrate temperature during deposition.

Lead on Copper-nitride

In order to investigate whether it is possible to manipulate the proximity effect present for Pb on pure Cu substrates additional experiments should be carried out on Pb deposited on an insulating layer of CuN. This system has not been described in the literature so far and thus, as a first step, the growth properties had to be examined. Fig. 6.10(a) shows a STM topography after depositing Pb for 75 s on the CuN/Cu(100) substrate which was still at a temperature of about $50\,°\mathrm{C}$ after the preparation of the CuN layer. Large areas of the surface exhibit irregularly shaped islands on which the rectangular trenches typical for CuN/Cu(100) are still clearly visible. Additionally, sometimes there are islands on top of the underlying structure that extend laterally to some 10 nm. Also on this islands the CuN characteristics are observable. In between the islands there are lower regions that are rather contaminated by some kind of adsorbate lumps and a careful look reveals the trenches of CuN also in these areas.

The preparation procedure without checking the quality of the CuN film was chosen in the end when the latter was already routinely mastered because it was found that depositing Pb on the substrate below room temperature after

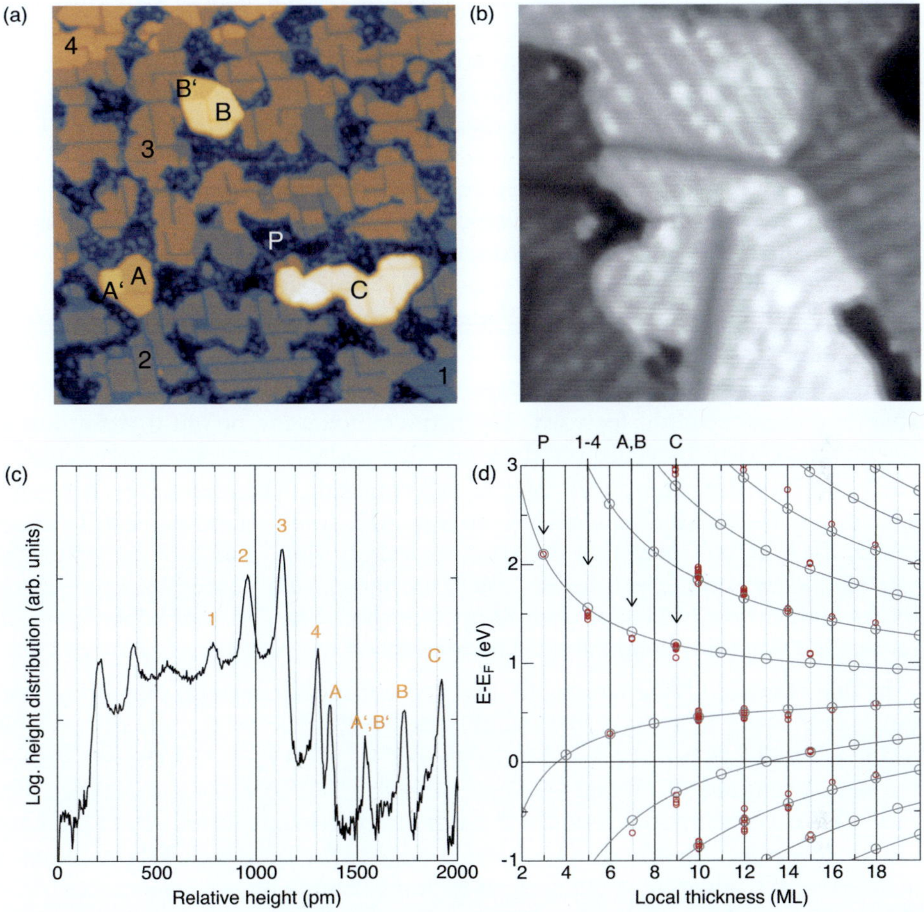

Figure 6.10: (a) STM topography ($150 \times 150\,nm^2$, 1 V, 100 pA) of Pb/CuN/Cu(100) after a deposition time of 75 s. The complicated appearance comprises as a main feature large islands of a unique thickness that are connected in many points and exhibit the prominent trenches stemming from the CuN layer below. The lowermost areas in between the islands are polluted with lumps. Only a few islands overtop the other structure. (b) Detail ($24 \times 24\,nm^2$) of some terraces revealing a stripe like structure in contrast to pure CuN which only shows trenches. (c) Height profile of (a): Four underlying Cu(100) steps can be identified. The determination of the local Pb thickness, in contrast, is not possible. (d) Matching of all observed QWS to the infinite square well model with a correction of δ_0=0.13 ML which allows to estimate the number of Pb layers.

taking it out from the cold STM, leads to the formation of films of homogeneous thickness. First, from a practical point of view, a sample exhibiting only one distinct Pb thickness does not allow to investigate possible dependencies on the number of layers in one measurement. Further, in the beginning, this sample morphology hindered the identification of Pb on the surface since, also for the totally closed films, the characteristic structure of the underlying CuN layer even shines through several layers of Pb. For this reason one could misleadingly assume from the first impression that there is no Pb at all present at the surface. Only on the second look one gets aware of an additional stripe like structure (Fig. 6.10(b)) which does not exist on pure CuN. In addition, STM images with atomic resolution revealed a (111) surface with the measured distances matching those of lead and therefore the surface is actually covered with (111) oriented Pb films.

Due to the rather complicated appearance the determination of the local Pb coverage is not straight forward. Fig. 6.10(c) shows the height distribution of the image in (a) and reveals that there are four underlying terraces separated by 1.8 Å and thus they stem from the Cu(100) substrate. The step edges, however, are quite deformed by the structure on top. Also most other height differences match the Cu(100) interlayer distance, especially A-A′ and B-B′ on top of the two smaller islands. It is this issue that hampers further attempts of extracting the local thickness of Pb from height profiles since one does not know to which terrace the apparent height should be referenced.

In this context, QWS, which are also present in this system, play an important role in order to gain further insight into the local coverage. As an example, islands A and B, which both exhibit a step of 1.8 Å on their top, indeed exhibit the same peaks at 1.25 eV and −0.72 eV in their dI/dU spectra all over their surface. This reveals that they consist throughout of the same number of Pb layers. On the lower island structure (1-4) the energy of the positive QWS increased a little while the negative is not present anymore which means, as expected, that the thickness is less while the opposite is true for island C. Interestingly, there is also a peak detected on the polluted lowermost area (P) at an energy of about 2.1 eV. Taking into account all the observed energies of QWS from this and from all other samples allows a matching to the infinite square well model presented in Fig. 6.10(d). An additional phase corresponding to an enhancement in addition to the actual thickness of $\delta_0 = 0.37\,\text{Å} = 0.13\,\text{ML}$ had to be assumed especially in order to cope with the states at around 600 meV which do not stay constant as in the pure model but are slightly decreasing with decreasing thickness. Again there is an oscillating bilayer stability which changes from odd to even around 10 ML. According to this "fit", the lower island structure in Fig. 6.10(a) has a local Pb thickness of 5 ML while the two smaller and the larger island consist of 7 and 9 Pb layers, respectively. Interestingly, also the peak position found on the dirty lowermost region fits the model quite reasonably, if a thickness of 3 ML is assumed meaning that, in fact, the whole sample surface seems to be covered by lead. Since the deposition rate was estimated to be in the order of 1 ML/min, this coverage seems to be too high, because the deposition time was only 75 s for this sample. This implies moreover, that also the determination of the other thicknesses might be off by 1 or 2 ML since there is quite some freedom in choosing the phase correction δ_0.

Even though, for this study, the approximate knowledge of the thickness is sufficient, in order to achieve final certainty about the local Pb coverage additional experiments with a more precisely determined total deposition rate are necessary. Furthermore, neither the reason for the presence of the lumps only on the supposed 3 ML region nor their chemical identity could be identified yet. Concerning the first, one might wonder whether again QWS play a role. Indeed, in a previous study it was observed experimentally that QWS have an influence on surface chemical reactivities [162].

The main purpose of introducing the additional CuN layer was to use its insulating property in order to retrieve superconductivity of the Pb structure on top. However, it turned out that the CuN layer is not capable of preventing the proximity effect and no sign of superconductivity was found even for the thickest obtained Pb layers of more than 20 ML.

6.4 Tunneling Spectroscopy on Lead on Silicon

In contrast to the situation on a metallic substrate, it is known from numerous studies that on silicon even very thin Pb islands of only a few atomic layers remain superconducting. In this section the results of the experiments addressing the phonon features in the tunneling spectra of Pb islands on Si(111) are presented, first in the superconducting state and afterwards in the normal state, after explaining the effect of Pb islands on the application of a magnetic field.

6.4.1 Phonon Features in the Superconducting State

The superconducting nature of Pb islands on silicon could be confirmed in this study as well. Fig. 6.11 shows a tunneling spectrum of a 16 ML thick island. For the measurements presented in Fig. 6.11 a superconducting tip was used, which resulted by chance/accident when the freshly prepared tungsten tip was unintentionally scratched across the sample surface during a lateral positioning step of the tip. Obviously, this procedure caused the tip to rip off one or several Pb islands from the sample surface leaving a superconducting Pb terminated tip. Due to the superconducting tip the quasiparticle peaks in the spectrum in Fig. 6.11(a) show up at about $\pm 3\,\mathrm{meV}$ corresponding to $2\Delta_{Pb}$ as expected from the theory of tunneling between two superconductors [7].

The most important feature in the spectrum is, however, the clear appearance of wiggles at the decaying edges of the quasiparticle peaks. These features turn up as two very prominent peak-dip pairs at $\pm 7\,\mathrm{meV}$ and $\pm 11.5\,\mathrm{meV}$ after zooming in on the $\mathrm{d}^2I/\mathrm{d}U^2$ spectrum as presented in Fig. 6.11(b) and unambiguously arise from the phonons of lead. Similar to the case of niobium before, according to Eliashberg theory the van Hove singularities at $4\,\mathrm{meV}$ and $8.5\,\mathrm{meV}$ in the phonon density of states of lead appear as dips/peaks at the positive/negative bias side in the electronic density of states shifted by the energy gap of the superconducting island. The additional shift of $1.5\,\mathrm{mV}$ is due to the superconducting Pb terminated tip.

On the one hand, this finding is not surprising because it is in absolute agreement with the early experiments on planar Pb-oxide-Pb tunnel junctions [9]. On the other hand, the verification that the same results can be obtained for a nano-sized sample in the STM geometry was still missing. In fact, other studies on Pb islands on silicon also observed hints of phonon features in the electronic density of states [60] but spectra with the energy resolution of the present work, which is due to the reduced temperature of less the 1 K and also partially to the superconducting tip, were by far not obtained before.

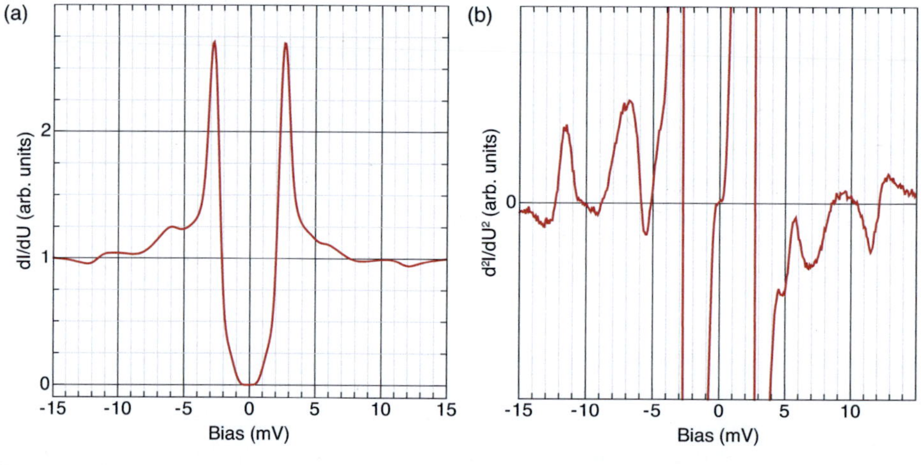

Figure 6.11: Tunneling spectroscopy on a 16 ML thick Pb island on Si(111): (a) $\mathrm{d}I/\mathrm{d}U$ spectrum normalized to the value at -15 mV which reveals the superconducting nature of the Pb island and shows wiggles at the decaying edges of the quasiparticle peaks. (b) The same data presented as $\mathrm{d}^2I/\mathrm{d}U^2$ and zoomed in on the low intensity region. Outside the gap there are two pairs of peaks and dips reflecting the Pb phonons in the electronic DOS of the superconducting Pb island. The spectrum in (b) was recorded using the 2f mode of the lock-in amplifier (U_{mod}=500 μV) at T=895 mK and (a) is the numerical integration. A superconducting Pb terminated W tip exhibiting more or less the same gap as the island was used for the measurement.

The data of Fig. 6.11 could now be used as starting point for the McMillan inversion algorithm in order to extract the Eliashberg function. However, regarding the main purpose of this work to detect the phonons of the material by an inelastic scattering process of the tunneling electrons in the normal state, in the next section, the possibility of suppressing superconductivity in the Pb islands on Si(111) by a magnetic field is addressed. All following results were achieved in another measurement session with a normal conducting tip.

6.4.2 Pb on Si(111) Exposed to a Magnetic Field

The result of applying a magnetic field is exemplified at a part of a quite large, irregularly shaped Pb island, the topography of which is presented in Fig. 6.12(a). Since the island is located on top of a very wide ($> 1\,\mu$m) Si terrace there are no underlying steps and the island thus possesses a unique height determined to be 10 ML including the WL. As expected from previous studies, the island is found to behave as a type-II superconductor with the magnetic flux penetrating in the form of vortices [111, 163].

Images (b) to (e) show $\mathrm{d}I/\mathrm{d}U$ maps for different magnetic fields, respectively. By setting the bias to $1.5\,\mathrm{meV}$ the energy corresponds to the quasiparticle peak thus allowing to distinguish superconducting areas that exhibit an enhanced density of states and appear bright from normal conducting ones, which have no quasiparticle and thus look darker. This becomes clear when regarding Fig. 6.12(g) showing tunneling spectroscopy of the superconducting gap at points getting closer to a vortex core as indicated in (b). It is worth mentioning that there is still a reduced conductivity at zero voltage at the measurement point closest to the center of the vortex, giving an impression that it is rather difficult to exactly locate the normal conducting core.

In contrast to a sample with a large flat surface the geometry of the islands prevents the evolution of a regular hexagonal vortex lattice. Rather, the individual vortices seem to be distributed in such a way that they find an optimal position inside the given shape of the island. For example, at $B = 0.23\,\mathrm{T}$, which was the smallest magnetic field applied to the sample, five vortices enter the island in the area as shown in Fig. 6.12(b), while there is one more at $0.30\,\mathrm{T}$ (Fig. 6.12(c)) and the positions of the vortices are redistributed a little. From a simple calculation one would expect about 14 and 18 vortices penetrating the displayed surface of the island at $0.23\,\mathrm{T}$ and $0.30\,\mathrm{T}$, respectively. While the actual number of vortices is significantly smaller, the distance of individual vortices is roughly matching the theoretical expectation as indicated by the red circles. After increasing the magnetic field up to $0.41\,\mathrm{T}$ (Fig. 6.12(d)), the number of vortices in the main part of the island again increases and the distance between them gets even smaller so that some of the vortices start to coalesce. An additional vortex now also appears in the arm at the upper left of the island.

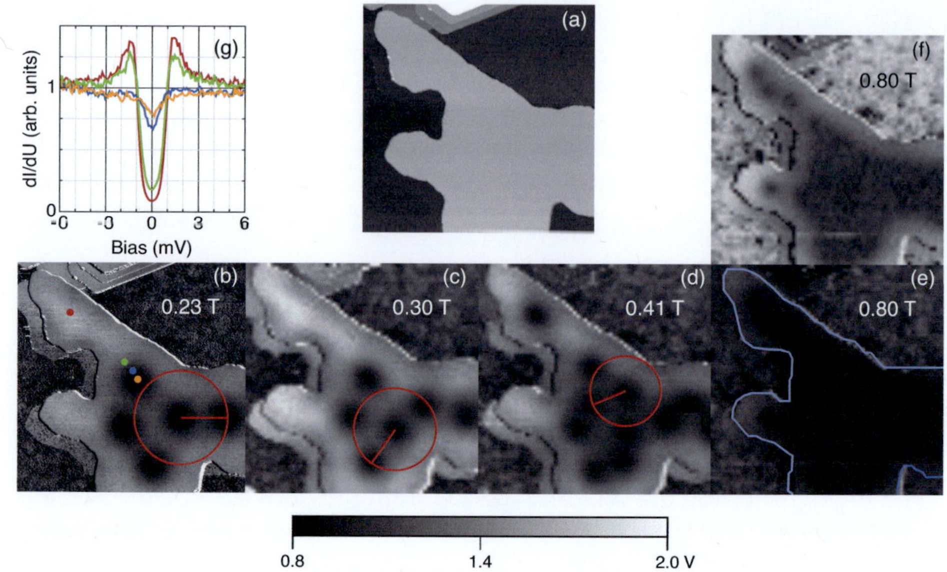

Figure 6.12: Type-II superconductivity of Pb islands on Si(111): (a) STM image ($500\times500\,\mathrm{nm}^2$, 1.5 mV, 100 pA, 0.75 K) of an irregularly shaped island exhibiting a height of 10 ML including the WL. (b) - (e) Maps of the lock-in signal proportional to $\mathrm{d}I/\mathrm{d}U$ with the bias set to the energy of the quasiparticle peak at 1.5 meV for different values of the magnetic field given in the images. The given color code is the same for images (b) - (e). (g) $\mathrm{d}I/\mathrm{d}U$ spectra taken at the points indicated in (b). The magnetic flux penetrates the island in the form of vortices for $B \leq 0.41\,\mathrm{T}$. While the arrangement of the vortices reflects the shape of the island, the distance between them corresponds roughly to the one expected from theory as indicated by the red circles. Though, at 0.8 T the island appears almost totally dark and superconductivity seems to be completely suppressed, the same map represented in a different color code in (f) reveals that there is remaining superconductivity at the island edge.

The vortex is located at the widest part of the arm providing most space, once more illustrating that the island shape has a distinct influence on the vortex positions. On the right side of Fig. 6.12(d) it seems that some vortices suddenly moved while the STM image was recorded indicating that redistribution of vortices eventually also takes place without changing the magnetic field probably due to more than one energetically favorable vortex positioning.

At $0.8\,\mathrm{T}$ the island appears more or less completely dark in the $\mathrm{d}I/\mathrm{d}U$ map shown in Fig. 6.12(e) in the same color code as the previous ones, which means that B_{c2}, the magnetic field above which the superconductivity is totally suppressed inside the material, has been exceeded now. Interestingly, even at this high value of the magnetic field, the edge of the island still appears brighter then the interior after adjusting the color code properly (Fig. 6.12(f)), indicating that superconductivity remains at the surface of this 2D object even above B_{c2}. This real-space observation of surface superconductivity confirms the experiments of Ning *et al.* [163], who investigated the same system by STM and also found persisting superconductivity at the island edges. Although there is a general agreement, the values of the critical fields differ quite substantially. The previous study determined $B_{c2} = 196\,\mathrm{mT}$, which accords well with the result of experiments using a superconducting quantum interference device magnetometer [164]. On the other hand, the present study reveals that B_{c2} must be at least $0.41\,\mathrm{T}$. Furthermore, Ning *et al.* obtained $308\,\mathrm{mT}$ for the additional critical field B_{c3}, above which also the surface superconductivity disappears,[5] while here, from Fig. 6.12(f), it is clear that there are remains of superconductivity at the island edge even at $B = 0.8\,\mathrm{T}$. It may be speculated whether the disagreement can be attributed to the higher temperature ($\approx 4.3\,\mathrm{K}$) in the previous studies while it was $0.75\,\mathrm{K}$ in the present case.

6.4.3 Phonon Features in the Normal State

From Fig. 6.12 it is clear that, when the Pb island is in the Shubnikov phase, superconductivity is fully suppressed only in the center of a vortex, which is rather difficult to locate exactly. The full superconducting density of states is reached within the coherence length. Any spectrum will thus be rapidly dominated by the superconducting gap when moving away from the core. For this reason first attempts to probe phonons by inelastic tunneling spectroscopy were made at magnetic fields $B > B_{c2}$ in the normal conducting interior of Pb islands.

Fig. 6.13 shows the result of such a measurement on a 11 ML thick island. The magnetic field was set to $B = 1.05\,\mathrm{T}$. Surprisingly, by far the most prominent feature in the $\mathrm{d}^2I/\mathrm{d}U^2$ spectrum (Fig. 6.13(a)) is still a peak close to zero voltage. While the I-U curve in Fig. 6.13(b) looks rather flat, as expected for the normal state, Fig. 6.13(c) reveals that the differential conductance indeed drops by $25\,\%$ at the Fermi level. Whether this feature is due to remains of superconductivity or originates from other processes could not be clarified yet. A second discovery is the appearance of a peak at $4\,\mathrm{mV}$, which shows up especially clear in the magnification in the inset of Fig. 6.13(a). The position of this peak would correspond exactly to the signal expected for the detection of the van Hove singularity of transversal phonons by inelastic scattering with the tunneling electrons. This interpretation, however, has to be treated with some caution. First, it is not obvious why the tunneling electrons should only probe the transversal modes while there is no sign of interaction with the longitudinal

[5] The existence of a superconducting sheath at the edge of a type-II superconductor for magnetic fields $B_{c2} < B < B_{c3}$ can be understood theoretically from the Ginzburg-Landau theory [165]. Within this framework a universal ratio $B_{c3}/B_{c2} = 1.69$ can be derived as well.

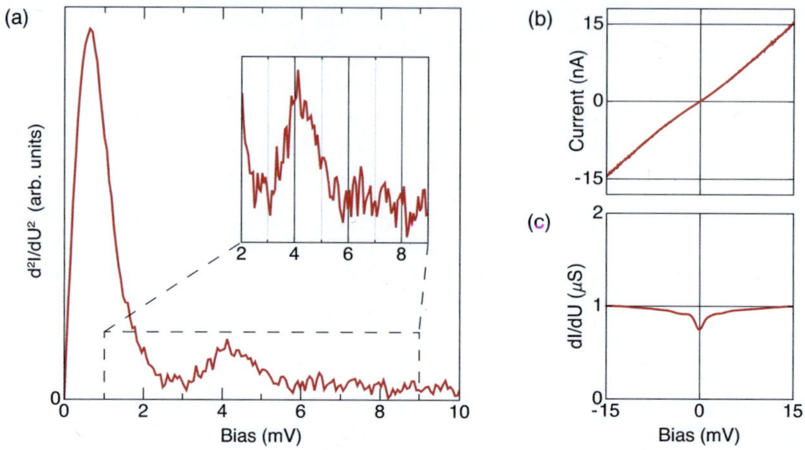

Figure 6.13: Inelastic tunneling spectroscopy at the normal conducting interior of a 11 ML thick Pb island on Si(111): (a) $\mathrm{d}^2I/\mathrm{d}U^2$ spectrum (average of the positive and negative bias side). The inset clearly reveals a peak at 4 mV. (b) I-U curve. (c) Differential conductance exhibiting a drop of 25 % at zero voltage.

ones. A second problem is that the disordered wetting layer does not allow to take proper reference spectra in order to check whether the observed signature is a feature of the tip. Due to the weak and doubtable signal, attempts to observe the phonon features at smaller fields in the center of a vortex remained without results although the general feasibility was demonstrated in the case of niobium in the previous chapter. An explanation for the unsatisfying outcome of this experiment could be the condition of the STM tip. It is well known that not every tip is well suited for doing inelastic tunneling spectroscopy and the experiments presented here stem from only one measurement session throughout performed with the same tip. In this respect, only a repetition of the experiments on Pb islands on silicon involving different tip conditions could give a definite answer to the question whether a direct measurement of the Eliashberg function of this system is possible or not. Fast retries, however, were hindered by the fact that the used silicon substrate lost its conductivity at low temperature when undergoing a second preparation cycle. In this situation, the study was quit for the lack of time and because experiments on Pb islands on normal conducting substrates appeared more promising with respect to the direct observation of the Eliashberg function as demonstrated in the next section.

6.5 IETS Results for Lead on Copper

6.5.1 Phonon Features and Eliashberg Function

As explained in detail before, Pb islands on copper are already in the normal state due to the inverse proximity effect. This is true for islands of all observed thicknesses on the two differently oriented surfaces as well as on the additionally introduced insulating layer of CuN/Cu(100) and supersedes the application of a magnetic field in order to reach the normal state.

First, Fig. 6.14 shows exemplarily the results of $\mathrm{d}I/\mathrm{d}U$ spectroscopy on a Pb island with a thickness of 13 ML on Cu(111) as well as on the wetting layer on Cu(111). In contrast to the constant differential conductance of the wetting layer there is a pronounced dip around E_F in the case of the 13 ML Pb island. The same effect was observed for Pb islands of various thickness on Si(111)-(7×7) [166]. Since that experiment was performed at 10 K, which is above T_c of Pb, the authors termed this feature "pseudo gap" as in the field of high temperature superconductivity.

In contrast, another explanation is given here by attributing the dip to inelastic excitations. Fig. 6.14(b) presents a high-resolution $\mathrm{d}^2I/\mathrm{d}U^2$ spectrum of the relevant energy range. An almost point-symmetric shape is obtained showing two dominant dip-peak pairs at a bias of about $4\,\mathrm{mV}$ and $8.5\,\mathrm{mV}$ as well as one very close to zero bias ($0.6\,\mathrm{mV}$). In order to obtain results of the present quality the tip was usually stabilized at $20\,\mathrm{mV}$ and a quite high current of $50\,\mathrm{nA}$. The lock-in amplifier was set to a modulation voltage of typically less than $500\,\mu\mathrm{V}$ at a frequency of $16.2\,\mathrm{kHz}$ to record a 2f signal which is thus affected only little by additional broadening. As a result, however, in order to achieve a sufficient

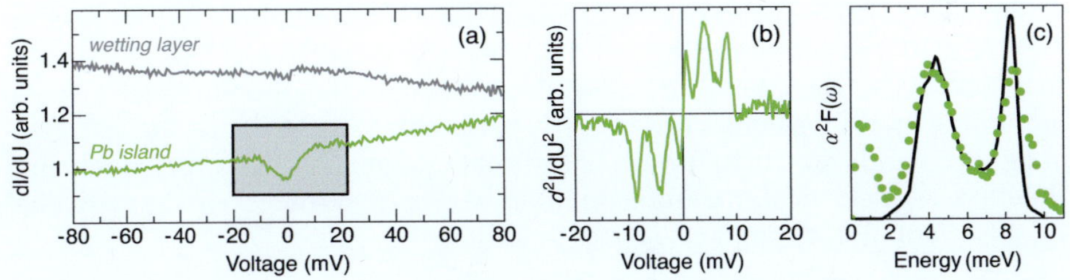

Figure 6.14: Tunneling spectroscopy on a 13 ML thick Pb island on Cu(111): (a) Large energy range tunneling conductance spectra (normalized and vertically shifted for clarity) on the WL and on the island. While the conductance of the WL is constant, it exhibits a significant decrease at zero bias on the Pb island. (b) A $\mathrm{d}^2I/\mathrm{d}U^2$ spectrum taken on the same position in a smaller bias interval around E_F. (c) Average of the positive and negative bias side of the IETS spectrum in (b) (dots) in comparison with the previous result for the Eliashberg function of Pb [9].

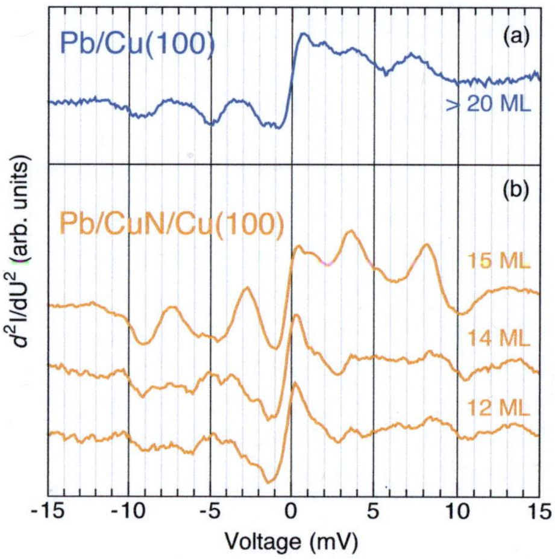

Figure 6.15: IETS on Pb/Cu(100) and on Pb/CuN/Cu(100): (a) $\mathrm{d}^2I/\mathrm{d}U^2$ spectrum of a Pb island on Cu(100) of more than 20 ML thickness (T=1 K, U_{mod}=0.3 mV). (b) $\mathrm{d}^2I/\mathrm{d}U^2$ spectra of Pb areas of the indicated local thicknesses on CuN/Cu(100) (T=1 K, U_{mod}=0.3 mV). While the phonon signature is clearly observable in the case of 15 ML it is almost not present for 12 and 14 ML.

signal-to-noise ratio, many single runs had to be averaged so that the total acquisition time for one spectrum was commonly as long as about one hour.

A comparison to the previously determined Eliashberg function of Pb [9] is provided in Fig. 6.14(c) and since it reveals an almost perfect agreement there is strong experimental evidence that IETS directly yields the Eliashberg function as was found by the theoretical calculation discussed in Chap. 3. Only the feature closest to E_F is not related to $\alpha^2F(\omega)$ but is due to a zero-bias anomaly (ZBA), as already observed in planar tunneling experiments on Pb [63], the origin of which has not been explained yet.

Similar $\mathrm{d}^2I/\mathrm{d}U^2$ spectra were obtained on Pb islands on Cu(100) and on CuN/Cu(100) as presented in Fig. 6.15(a) and (b), respectively. Due to the large thickness of the islands on Cu(100) no dependence of the phonon spectra on the amount of layers was observed in the experiments. Furthermore, the $\langle 100 \rangle$ oriented Pb islands on Cu(100) were unfortunately only observed very rarely and therefore no inelastic tunneling data could be recorded there so that a statement on the orientation dependence is not possible.

In the case of Pb on CuN/Cu(100) there were hints for an influence of the local thickness on the intensity of the phonon related features in the tunneling spectra. The three measurements presented in 6.15(b) were successively performed with the same tip conditions on Pb areas of the indicated thicknesses. While the phonon features at ± 4 mV and ± 8.5 mV are very prominent for 15 ML, they almost disappear for 12 and 14 ML. Because the topography of Pb islands on Cu(111) can be interpreted much easier, this effect could be studied in more detail on this substrate as explained in the next section.

6.5.2 Thickness-Dependent Eliashberg Function of Pb/Cu(111)

After confirming that STM-IETS is able to measure the Eliashberg function the dependence of the electron-phonon coupling on the local Pb thickness was examined. In particular, the "wedge"-like island of Fig. 6.6 was investigated. Keeping all parameters constant, d^2I/dU^2 spectra were recorded on every local thickness. As shown in Fig. 6.16(a), the zero-bias anomaly and the Eliashberg function were found in all cases. While the former stays the same there is a strong thickness dependence of the intensity of $\alpha^2F(\omega)$. For areas comprising an even number of layers it is relatively weak whereas the signal gets stronger for 13 ML and is by far strongest for 11 ML as can also be seen in Fig. 6.16(b) which shows a d^2I/dU^2 map, i. e., a map of the local electron-phonon coupling at 8 meV, the energy of the longitudinal peak. No differences were found for different locations within an area of a given thickness. In all experiments, the current was set to the same value during tip stabilization, such that the differences in the inelastic spectra are not related to differences in the electronic density of states. Since an influence of the electronic QWS on the thickness dependence can be expected, the first derivative of the tunneling current in a larger voltage range was measured and is depicted in Fig. 6.16(c). For 12 and 14 ML, the highest occupied as well as the lowest unoccupied states lie roughly 0.5 eV away from the Fermi level. In contrast, the lowest unoccupied QWS of the 11 ML and 13 ML thick layers are very close to E_F. The relevant QWS of the former, which exhibited also the highest intensity in $\alpha^2F(\omega)$, is closest to E_F. Thus, this experiment clearly shows, that electron-phonon coupling can be strongly enhanced (factor 5), if an unoccupied QWS is near the Fermi energy.

In order to test this finding another sample with a total nominal coverage of 16 ML was prepared which allowed access to local thicknesses between 17 ML and 33 ML. In this thickness range the Fermi level is again crossed by a QWS at 18 ML and 27 ML as was seen in Fig. 6.8(b). The spectroscopic results for Pb thicknesses around these two numbers of monolayers are presented in Fig. 6.17. In contrast to the lower coverage, where a thickness of 9 ML was not observed at all, now also those two important thicknesses for which a QWS is located directly at the Fermi level (Fig. 6.17(a) and (d)) could be found. The influence of the QWS on the growth thus seems to be diminished as the Pb slab becomes thicker.

Fig. 6.17(b) and (e) show dI/dU spectra recorded in a narrower voltage range. As expected, they all reveal a dip at zero bias[6], especially, though not so pronounced, also for 18 and 27 ML. At these thicknesses Wang *et al.* [166] claimed

[6]The very asymmetric shape of the 1f spectrum on 20 ML around zero voltage (red curve) reflects the onset of a QWS just above the Fermi energy. This feature also propagates into the 2f spectrum and therefore only the mirrored negative bias part, instead of the antisymmetrized data, is shown in 6.17(c).

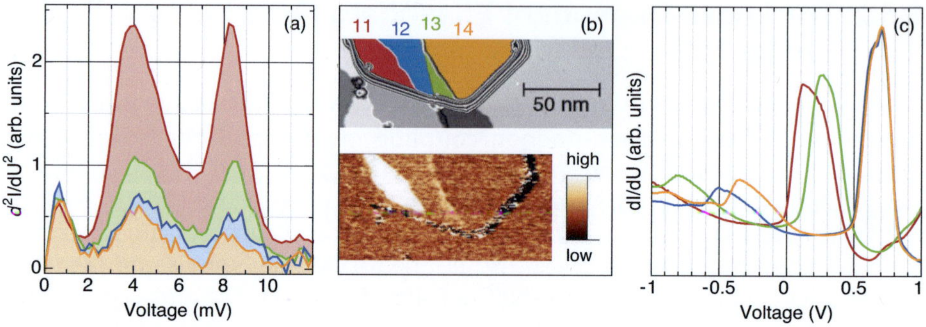

Figure 6.16: Thickness dependent electron-phonon coupling: (a) Antisymmetrized $\mathrm{d}^2I/\mathrm{d}U^2$ spectra recorded at an arbitrary location on the areas of different local Pb coverage as indicated by the colors in the upper topographic image of the wedge-shaped island in (b). The intensity of $\alpha^2F(\omega)$ increases considerably on the 11 ML high Pb area as can also be seen from the $\mathrm{d}^2I/\mathrm{d}U^2$ map of the island taken at 8mV which is depicted in the lower part of (b). (c) QWS obtained from $\mathrm{d}I/\mathrm{d}U$ spectra at the same locations as the spectra in (a).

a "pseudo peak" for Pb on Si(111) and explained this finding by a phenomenological model, borrowed from optics. This model, however, is thus not able to describe the observation in general and the interpretation by inelastic phonon excitation appears much more trustworthy, especially when regarding the $\mathrm{d}^2I/\mathrm{d}U^2$ spectra presented in Fig. 6.17(c) and (f).

These spectra reveal the same fine structure as for the sample of lower coverage. Also the correlation of intensity and QWS position could be confirmed. At 20 ML the QWS, which is located directly at the Fermi level for 18 ML, has just shifted above E_F and is now unoccupied. In analogy to the behavior observed before on 11 ML, this is accompanied by an enhanced electron-phonon coupling as can be seen from the comparison of the $\mathrm{d}^2I/\mathrm{d}U^2$ spectra in Fig. 6.17(c).

Interestingly, when the QWS lies directly at E_F for 18 ML, this does not result in a further increase of intensity but the spectrum deviates substantially from those on the other thicknesses. The same is true for 27 ML, though in this case at least the "transversal" peak is still observable but also with reduced intensity[7].

To elucidate the experimental finding of an enhancement of the electron-phonon coupling when an unoccupied QWS is close to the Fermi energy, the electronic structure and the phonons for different free standing Pb films were calculated [167]. The structural relaxations and the phonons were computed

[7]One has to note that the intensities of the spectra from 6.17(c) and (f) can unfortunately not be compared due to a higher experimental temperature and very different tip conditions (f). In the latter case the tip produced very asymmetric spectra so that only the negative bias side showed meaningful signatures. Only this part is presented (after mirroring) in 6.17(f).

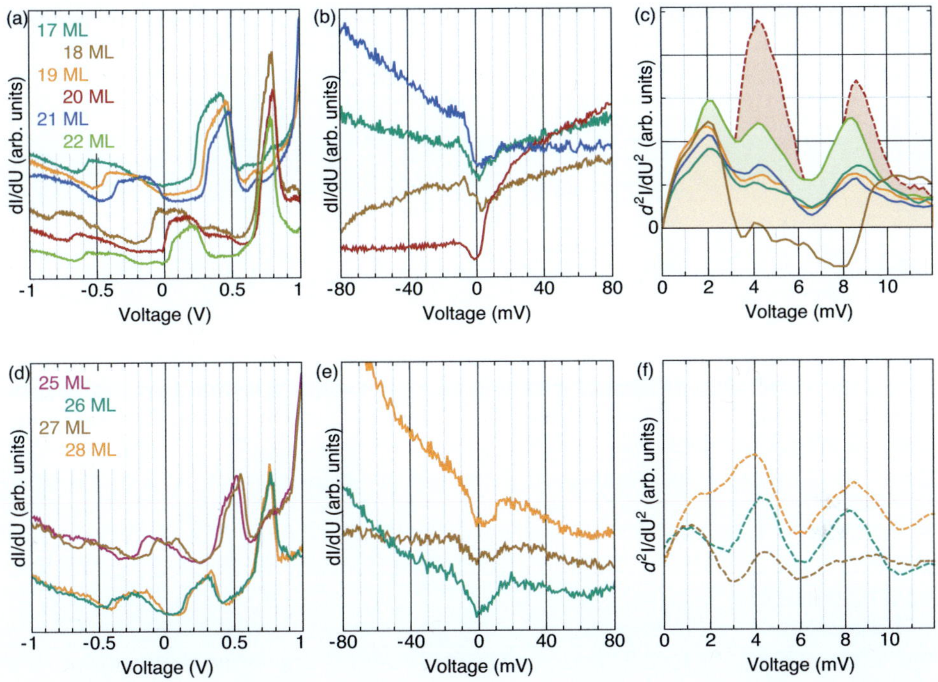

Figure 6.17: Spectroscopic results for thicker Pb films: (a) and (d) show the QWS in $\mathrm{d}I/\mathrm{d}U$ spectra for the indicated thicknesses of 17 to 22 ML and 25 to 28 ML, respectively. (b) and (e) display $\mathrm{d}I/\mathrm{d}U$ spectra in a narrower voltage range revealing dips at zero bias on all measured thicknesses. (c) and (f) show the corresponding antisymmetrized $\mathrm{d}^2I/\mathrm{d}U^2$ spectra (The dashed lines indicate spectra where, due to different reasons, only the (mirrored) negative voltage part is shown (See footnotes). The missing curves could not be measured because of a dramatic tip change before the series was completed).

using the VASP code, well known for precise total energy and forces calculations [168], while the electronic structure of the thin films was obtained with a first-principles Green's function method, specially designed for semi-infinite systems such as surfaces and interfaces [169]. The self-consistently calculated Green's functions and phonons were used to compute the Eliashberg function of the given systems. First of all, QWS were found in these systems in qualitative agreement with the experiment. The calculated electronic density of states exhibits a QWS closest to the Fermi level for a thickness of 12ML while for 10 ML and 14 ML the QWS lie roughly 0.1 eV below and above E_F, respectively (Fig. 6.18(a)). The theoretical results for $\alpha^2F(\omega)$ in Fig. 6.18(b) reveal the same effect as the experiment: While the intensity of $\alpha^2F(\omega)$ is the same for 10 and 14 ML, it increases by more than 30 % in the case of 12 ML. Also shown in Fig. 6.18

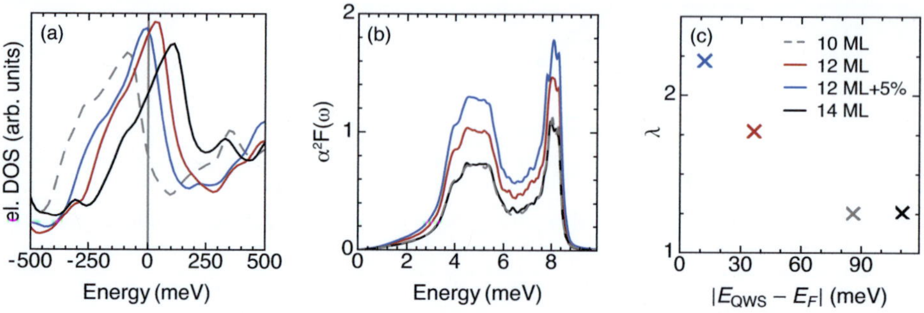

Figure 6.18: Theoretical calculation [167] for free-standing Pb slabs with a thickness of 10, 12, and 14 ML: (a) QWS in the electronic DOS (at the first vacuum layer). (b) Eliashberg function. (c) Resulting electron-phonon coupling constant in dependence of the distance of the QWS from E_F.

is the effect of compressing the surface of the 12ML slab by 5 %. This compression shifts the QWS even closer to E_F which again is accompanied by a further enhancement of $\alpha^2F(\omega)$.

The resulting electron-phonon coupling constants λ, which change from a value of about 1.25 to 2.2, i. e., significantly above the bulk value, upon the shift of the QWS towards E_F are indicated in Fig. 6.18(c). Thus, in qualitative agreement with the experiment, the Eliashberg function can be increased by a QWS near the Fermi level. Quantitatively, the experimental enhancement of the electron-phonon coupling is even stronger than theoretically predicted by a factor of 2-3, possibly due to the details of the QWS of Pb on Cu(111).

An oscillating coupling between electrons and phonons in dependence of the Pb film thickness was also observed in an angle-resolved photoemission study where λ was extracted from the width of QWS peaks [120]. Crucial thicknesses (e. g. 18 ML), however, were not observed in this study. From the theoretical side, Brun *et al.* already showed in their ab-initio calculations of free-standing Pb films between 4 and 10 ML an influence of the density of states at E_F on the electron-phonon coupling constant [60]. An improved calculation by some of these authors, which includes spin-orbit coupling, also reveals the oscillation of λ with the electronic density of states [170].

The mechanism leading to the enhancement can be understood from total energy considerations. The formation of QWS in thin Pb films leads to an energetically unfavorable condition when a QWS lies very close to E_F. A change of the lattice constant, e. g. by a phonon, shifts the QWS up and down through the Fermi level and leads to repopulation of the electrons with large changes of the electronic energy. Thus, deformations couple more strongly to the electronic degrees of freedom when a QWS is near the Fermi energy, thereby increasing the electron-phonon coupling.

In the present experiment, however, this mechanism seems to be present only for unoccupied QWS, i. e. QWS above the Fermi level, while no significant enhancement is found when an occupied QWS lies close to E_F. In the calculation, on the other hand, also a QWS very slightly below the Fermi level results in an enhancement of λ, which happens upon the 5 % compression of the surface. In the cases where a QWS state is located directly at E_F the experiment revealed a deviating behavior and it might be speculated whether an unexplained additional effect is responsible for this.

6.6 Summary

The electron-phonon coupling in lead was studied after confirming its growth mode on Cu(100), Cu(111) and Si(111) which is characterized by islands on a wetting layer. The island heights are governed by electronic quantum well states. This is also the case for Pb on a closed film of copper-nitride on Cu(100), a system which has not been reported on in the literature yet.

Concerning the examination of phonon excitation, $\mathrm{d}^2I/\mathrm{d}U^2$ spectra of high quality could be obtained in the superconducting state on Si(111) as well as in the normal state, achieved by the inverse proximity effect, on Cu(100) and Cu(111). The same is true for Pb/CuN/Cu(100), where the additionally introduced insulating layer is not able to decouple Pb electronically from the copper substrate and thus superconductivity is still suppressed by the inverse proximity effect. While, for a unknown reason, inelastic tunneling spectroscopy on normal conducting Pb on Si(111), after suppression of superconductivity by a magnetic field, only revealed parts of the phonon characteristics, the $\mathrm{d}^2I/\mathrm{d}U^2$ spectra obtained on the other substrates can be unambiguously identified as $\alpha^2F(\omega)$ which allows for the experimental determination of the latter with the high spatial resolution of the STM.

For Pb islands on Cu(111) and partially also for Pb/CuN/Cu(100) a pronounced dependence of the Eliashberg function on the thickness of the Pb slabs was found which, with the aid of *ab initio* calculations, can be explained by their energetic stability due to the position of the electronic QWS.

7 First Experiments on Iron-Based Superconductors with the JT-STM

Having established in the two previous chapters that tunneling spectroscopy using the STM provides a precise tool in order to reveal the underlying boson structure in conventional superconductors it is now, of course, desirable to adapt these concepts to unconventional superconductors as well. For these the mechanism causing superconductivity is still under debate, though spin fluctuations seem to be a hot candidate, especially in the case of iron-based superconductors which were only discovered in 2008. Due to their complicated structure, however, unconventional materials in general have to be prepared in a very different manner compared to the single-element conventional ones. This chapter describes the necessary steps in order to obtain suited samples for experiments with the JT-STM. In the end it presents a test experiment on $Ba(Fe_{0.94}Co_{0.06})_2As_2$ *and first results on* $SrFe_2(As_{0.65}P_{0.35})_2$*, an iron-based superconductor that has not been studied by STM yet. In the beginning, however, information about iron-based superconductors is provided as far as it is needed to relate the preparation steps and the experimental findings.*

7.1 Motivation

7.1.1 Existing STM Experiments

Using tunneling spectroscopy to examine unconventional superconductivity is only possible thanks to the invention of the STM. After the discovery of high-temperature superconductivity in the cuprates, tunneling experiments using other geometries were not found to yield reproducible results [68].

Since, scanning tunneling microscopy and spectroscopy have been extensively applied to cuprates as well as to iron-based superconductors [32, 68]. In the meantime, STM experiments were carried out to investigate the surface morphology and the superconducting gap together with its possible spatial variation. The studies also cover more advanced issues like effects of impurities and imaging of vortex matter.

A rather sophisticated technique which was introduced by Hoffman *et al.* [35, 171] yields information about the Fermi surface and the momentum dependence of the order parameter $\Delta(\vec{k})$ from energy-dependent imaging of an interference pattern. This pattern appears on a sample surface with impurities when scattered quasiparticles interfere. Therefore the method is referred to as quasiparticle interference (QPI) or FT-STS, due to the 2D Fourier transform involved in the analysis. This kind of experiments complements angle-resolved photoemission (ARPES) studies that are usually carried out to obtain momentum-resolved data. The QPI technique has also been already applied to iron-based superconductors [172–175].

Most interesting with respect to the focus of the present study is the observation of so-called dip-hump structures outside the superconducting gap in iron pnictides [70] as well as in cuprates [68], as already discussed in Sec. 3.4.3. Due to the vivid current interest, the experiments described in the following focused only on iron-based systems.

7.1.2 Properties of Iron-Based Superconductors

The common building block of all known iron-based superconductors is a FeX layer, where X denotes a pnictogen (element of the nitrogen group) or a chalcogen (element of the oxygen group) [32]. The FeX layers may be stacked directly on top of each other or sandwiched between additional ionic layers. Corresponding 3D sketches of the crystal structure are depicted in Fig. 7.1(a) for the "11" and the "122" materials which are two examples of about six different families of iron-based superconductors [29]. While the parent compounds FeTe and $BaFe_2As_2$ (or $SrFe_2As_2$) do not show superconductivity it can be introduced by electron or hole doping as well as by introducing chemical pressure via isovalent doping with the substitutions indicated in the figure by brackets in the formulas of the respective compound.

A common feature of iron-based superconductors is the evolution of a superconducting dome next to a magnetically ordered phase in the T-x diagram [32]. The doping and the optimal level is given in Tab. 7.1 together with the corresponding T_c. All investigated samples were the optimally doped representatives.

compound	family	doping	x_{opt}	T_c	abbrev.	ref.
$Fe(Te_{1-x}Se_x)$	11	isovalent	$\approx 50\%$	14.5 K	FeTe-Se50	[176]
$Ba_{1-x}K_xFe_2As_2$	122	hole	40%	38 K	Ba122-K40	[177]
$Ba_2(Fe_{1-x}Co_x)_2As_2$	122	electron	$\approx 6\%$	23 K	Ba122-Co6	[178]
$Sr_2Fe_2(As_{1-x}P_x)_2$	122	isovalent	$\approx 35\%$	33 K	Sr122-P35	[179]

Table 7.1: Summary of the optimal doped iron-based superconductors considered in the present work.

The names of the families arise from their respective chemical composition. A "11" material without additional layers, e. g. FeTe, comprises the same number of Fe and of Te atoms. With respect to the unit cell, however, it should be rather referred to as "22" material since half of the Te atoms lie above the Fe plane while the other half lies below which results in a larger unit cell. This issue is illustrated in the top view of the FeX layer in Fig. 7.1(b). The one-Fe unit cell that would exist if all As were coplanar has a lattice constant $a_{\text{Fe-Fe}}$ of approximately 2.6 Å while the correct one is $a_0 = 3.9$ Å.

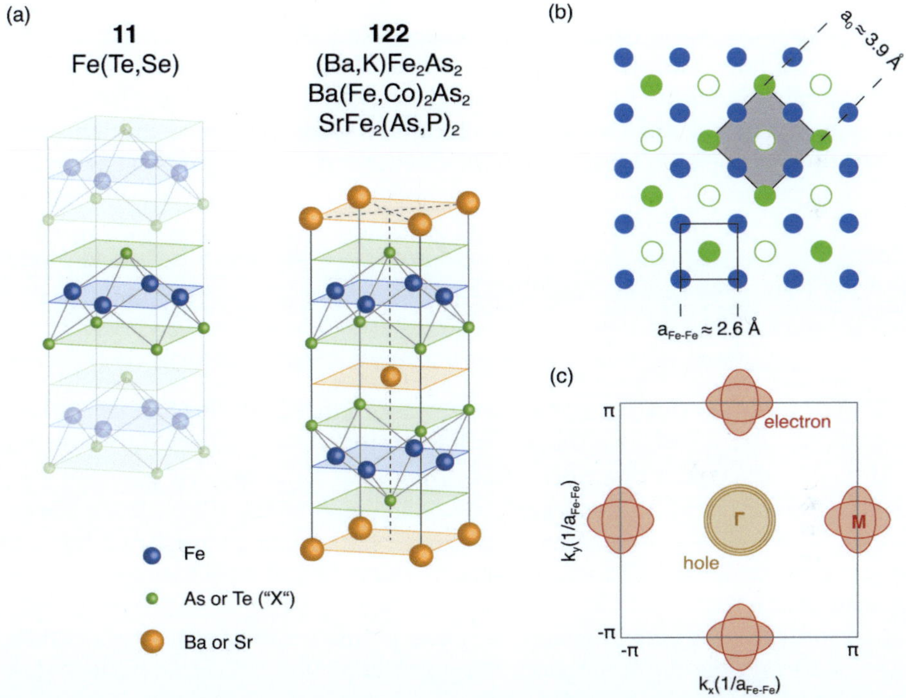

Figure 7.1: Structural and elecronic properties of iron-based superconductors: (a) 3D sketches of the parent compound crystal structures of "11" and "122" materials. The common building block responsible for superconductivity is the FeX layer (X = As or Se). Substitutions for electron, hole or isovalent doping may take place at arbitrary positions. The respective representatives of each family considered in this study are indicated. Adapted from [29]. (b) Top view of the FeX layer indicating different vertical positions of the X atoms by open and filled circles. Depending on whether the latter is considered or not a larger or smaller unit cell has to be drawn, respectively. Adapted from [32]. (c) Electron and hole pockets of the Fermi surface in the Brillouin zone (for one Fe atom in the unit cell) [29].

The choice of the unit cell also has consequences on the Brillouin zone in reciprocal space. The one depicted in Fig. 7.1(c) corresponds to the smaller unit cell in (b) neglecting the vertical position of the X atom. The complicated Fermi surface of iron-based superconductors, which is also sketched in the figure, is characterized by five sets of Fermi sheets arising from the d-orbitals of iron. Two of them form electron-like pockets at the M point, while the other three build up hole-like pockets centered at Γ [29]. Since these considerations are only based on the FeX layer, they apply for all iron-based superconductors. The different Fermi surfaces give rise to different gap values, characteristic for multi-band superconductors. Hints of multiple gap values have for example been observed experimentally in ARPES studies on Ba122-K40 [180] as well as in STM experiments on Ba122-Co6 [172] (see also [32]).

Very often, however, an even simpler sketch of the Fermi surface is employed only depicting one kind of electron and hole pocket, respectively. The electron and hole-like Fermi surfaces are linked via scattering processes involving the exchange of an antiferromagnetic spin fluctuation. In conjunction with this scattering process, if it was the mechanism for Cooper pair formation, an unconventional $s^{\pm}$ symmetry of the order parameter was proposed [181] which exhibits different sign on the electron and hole pockets. In this extended model the order parameter can be described by [182]

$$\Delta_{s\pm}(\vec{k}) = \Delta_1 \cos k_x \cos k_y + \Delta_2(\cos k_x + \cos k_y). \tag{7.1}$$

A 3D plot of this order parameter is shown in Fig. 7.2(a). It changes its sign on a nodal line, which is almost circular, centered at Γ. Depending on the ratio Δ_1/Δ_2 the nodal line of the full function $\Delta_{s\pm}(\vec{k})$ may or may not pass through the electron-like Fermi surfaces as depicted in Fig. 7.2(b). From the experimental side the $s^{\pm}$ symmetry of the order parameter is supported by STM experiments exploiting QPI in connection with an additional magnetic field in Fe(Se,Te) [175].

If the order parameter vanishes on some points on the Fermi surface then, for some directions, the superconducting gap vanishes. The quasiparticle density of states, which is averaged over all momenta, is thus not fully gapped anymore but, instead of a cup-shaped structure with steep edges, a V-shape results. While, due to symmetry reasons this is always observed in cuprates, which possess d-wave symmetry, in the pnictides the appearance of a V-shaped gap depends on the details of the compound under investigation as seen above. This effect was already observed experimentally in the comparison of tunneling spectra of $FeTe_{1-x}Se_x$, which are fully gapped [175] and FeSe, where a V-shaped gap was observed [182].

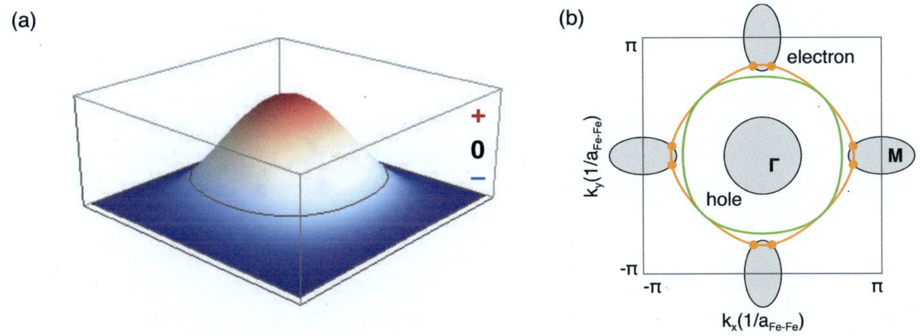

Figure 7.2: Superconducting order parameter for the $s^{\pm}$ symmetry: (a) 3D plot of the order parameter in the s± scenario for $\Delta_1/\Delta_2 = 1$ (b) Nodal line of the order parameter in the Brillouin zone. While it lies between the electron and hole pockets for $\Delta_1/\Delta_2 = 1$ (green line) the nodal line intercepts the electron-like part of the Fermi surface if Δ_1/Δ_2 is decreased (orange line). Orange dots mark the important points where the gap size on the Fermi surface is zero. Adapted from [32].

7.2 Sample Preparation

As explained in the previous section iron-based superconductors possess a rather complicated crystal structure which hampers an *in situ* growth by MBE of the respective samples. Indeed, there are such processes, e. g. in order to fabricate cuprate-based high temperature superconducting wires. Furthermore, it was shown, that FeSe and even KFe_2Se_2) crystals of very high quality can be grown by co-deposition of the materials on SiC(0001) [182, 183]. However, as soon as the chemical composition gets more complicated, more advanced setups are required in order to be able to survey and precisely control the exact ratio of the constituents [184].

For this reason the usual way is to obtain the bulk samples synthesized *ex situ* from groups who are specialized on the crystal growth and have the necessary experience to determine the appropriate recipes, see e. g. [179, 185]. The available single crystal samples are thus rare and valuable items. In order to arrive at sample surfaces suitable for STM investigation the common sputtering and annealing procedure cannot be applied. Instead the crystals are cleaved directly in the UHV in order to reveal a fresh surface without contamination from having been exposed to ambient conditions [32]. This procedure is possible due to the layered crystal structure of iron-based superconductors as presented in the previous section. However, it was found that the cleaving is much easier for the "11" family. Here the cleaving can be done by a simple "Scotch tape" method, as it is also applied in the case of graphite, while the layers of the "122" family are bound more strongly. For this reason, it is necessary to fix a post on the top side of the sample as illustrated in Fig. 7.3(a) by a two-component epoxy glue which provides a stronger cohesion to the crystal surface than the bonding between the layers.

However, due to the small sample size of typically around 1 mm^2, an example of which can be seen in 7.3(b), the positioning of the top post can hardly be accomplished by hand. Instead, a home-made instrument (Fig. 7.3(c)) came into use which allows to precisely position the post on top of the sample (d). The accurate control is especially needed in z-direction since, when the post is brought too close to the sample surface, a spillover of the glue would result in a stabilization at the side of the crystal, thus preventing it from being cleaved properly. After positioning the top post the whole setup can be placed on a hot plate in order to harden the glue. Subsequent to the transfer of the sample to the UHV system it is simply cleaved by pushing the wobble stick against the top post, which should point to the bottom such that it falls down and does not affect the freshly created surface.

At the JT-STM cleavage can be done at room temperature in the preparation chamber or at about 77 K at the pre-cooling state of the STM. Recovery of the second part of the crystal is accomplished by using a magnetic material for the top post[1] that is caught by a NdFeB magnet mounted on an additional sample plate which is placed in an empty slot of the pre-cooling stage some centimeters below the sample when it is being cleaved. Besides the fact that this part of the crystal can be used for later experiments, the recovery provides the possibility of judging the cleaving process since there is no convenient position at which the sample itself can be easily looked at. It was found that, after introducing the apparatus for positioning the top post, the rate of successful cleaving procedures was increased to 100%. Figure 7.3(e) shows a sample and the corresponding top post carrying the other half after they had been taken out from the UHV system after the measurement had been finished.

The small size of the sample in combination with the limited optical access and the lack of freedom of movement in one of the two lateral directions hampers the coarse approach of the STM tip to the sample surface. One has to bear in mind that missing the target only a little results in a contamination of the STM tip with epoxy glue which means the end of the experiment.

Concerning the focus of the present work, the study of the optimally electron-doped compound Ba122-K40 is certainly most promising since in this material very clear features above the superconducting energy gap were discovered in the spectra measured by STS [71, 72] as well as by ARPES [74] experiments and bosonic modes were proposed as origin. In this respect this compound would be a well suited candidate to first test these findings and subsequently study the appearance of the bosonic mode in the normal state similar to the experiments on conventional superconductors.

[1] In this study ordinary nails were used.

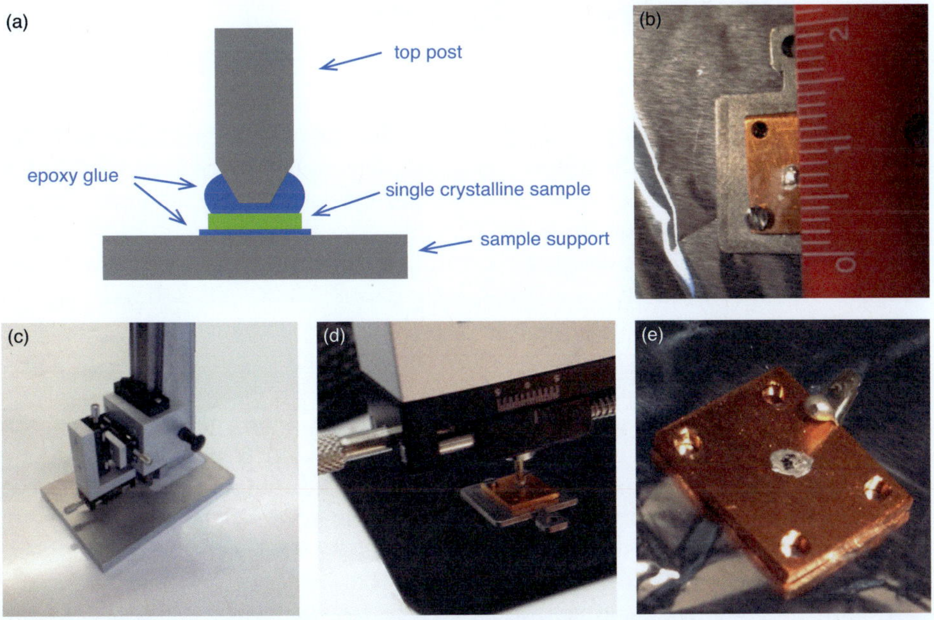

Figure 7.3: Preparation for UHV cleaving of the sample: (a) Sketch of the attachment of the top post for cleaving. (b) Photograph of a sample giving an impression of the typical sample size. (c) Home-made gadget for precise positioning of the top post as shown in (d). (e) Sample and recovered top post after the experiment.

There is, however, a problem concerning the handling of this compound due to the volatile nature of the comprised potassium. At ambient conditions it reacts with water from the air moisture leading to a chemical degradation of the sample. For this reason storage and transport have to take place under vacuum and sample preparation (gluing the top post and especially the subsequent hardening) has to be carried out under inert gas atmosphere in a glove box.

The corresponding experiments were done during a research visit at the group of F. Komori at the Institute for Solid State Physics at the University of Tokyo, where the necessary equipment was available, though in the end attempts of STM measurements remained without success because the cleaving process did not work out as required. At this time the gadget described above had not been at hand yet.

For this reason, the experimental results presented in the following section are restricted to FeTe-Se50, Ba122-Co6, and Sr122-P35 which can be handled and prepared at ambient conditions.

7.3 First Experimental Results

Topography

Figure 7.4(a) shows an atomically resolved STM image of the surface of FeTe-Se50 which was obtained in Tokyo[2]. The sample was simply cleaved by "Scotch tape" in UHV. The brighter spots in the apparent height were attributed to excess iron atoms in previous studies [175, 186].

This tidy surface morphology can be found in the case of cleaved 11 materials because the individual FeX layers can be separated in a well defined way. In contrast, iron pnictide compounds of the 122 family exhibit a poorer surface morphology after cleaving. Though they are expected to cleave with the FeAs layers intact as well, the structure on the surface is obscured by the intermediate buffer layer remaining partially on the one surface or on the other. This fact leads to the observation of many different surface morphologies depending on the details of the chemical composition of the sample and on the cleaving parameters. Besides many examples where a disordered surface was found, (2×1) and ($\sqrt{2}\times\sqrt{2}$) reconstructed surfaces were frequently observed, supposed to originate from 50 % of the intermediate layer atoms [32].

In agreement with previous studies [186] a test experiment on Ba122-Co6, which had been cleaved at room temperature, revealed a surface exhibiting the (2×1) reconstruction in domains of the two possible orientations (Fig. 7.4(b)). The spacing between the lines is 7.3 Å which deviates from $2a_0 = 7.8$ Å by 6.4 %, which is not uncommon for a STM study. This is due to the calibration of the scanner piezo, which depends on the image size and the scanning speed because of piezo creep. Atomic resolution of the shorter distance along the lines could not be achieved. Even this, in principle, well structured surface, however, is characterized by a lot of adsorbates and vacancies of unknown origin.

Topographic images of the surface of a Sr122-P35 sample, which were cleaved at 77 K are presented in Fig. 7.4(c)-(f). The sample was provided by the group of S. Tajima [179] and has not been studied by STM yet. Already on the large scale (Fig. 7.4(c) and (d)) the observed surfaces look different and the respective closeups in (e) and (f) also reveal a different atomic arrangement. The stripe-like structure in (e) exhibits a distance of 6.8 Å, deviating a little more from $2a_0$ (12.8 %) but is still supposed to be the (2×1) reconstruction. The surface in (f) appears rather blurred on the first sight. The 2D Fourier analysis in the inset, however, reveals a regular square pattern with a lattice constant of 4.9 Å, which deviates by 10.9 % from the one expected for the ($\sqrt{2}\times\sqrt{2}$) reconstruction. Interestingly, the two morphologies were found on one and the same sample in

[2]Surprisingly, no indication of superconductivity could be found in STS on the FeTe-Se50 sample. It could not be figured out whether the aged sample had been degraded already or the energy resolution and/or the base temperature of the used STM did not conform the specifications of the manufacturer.

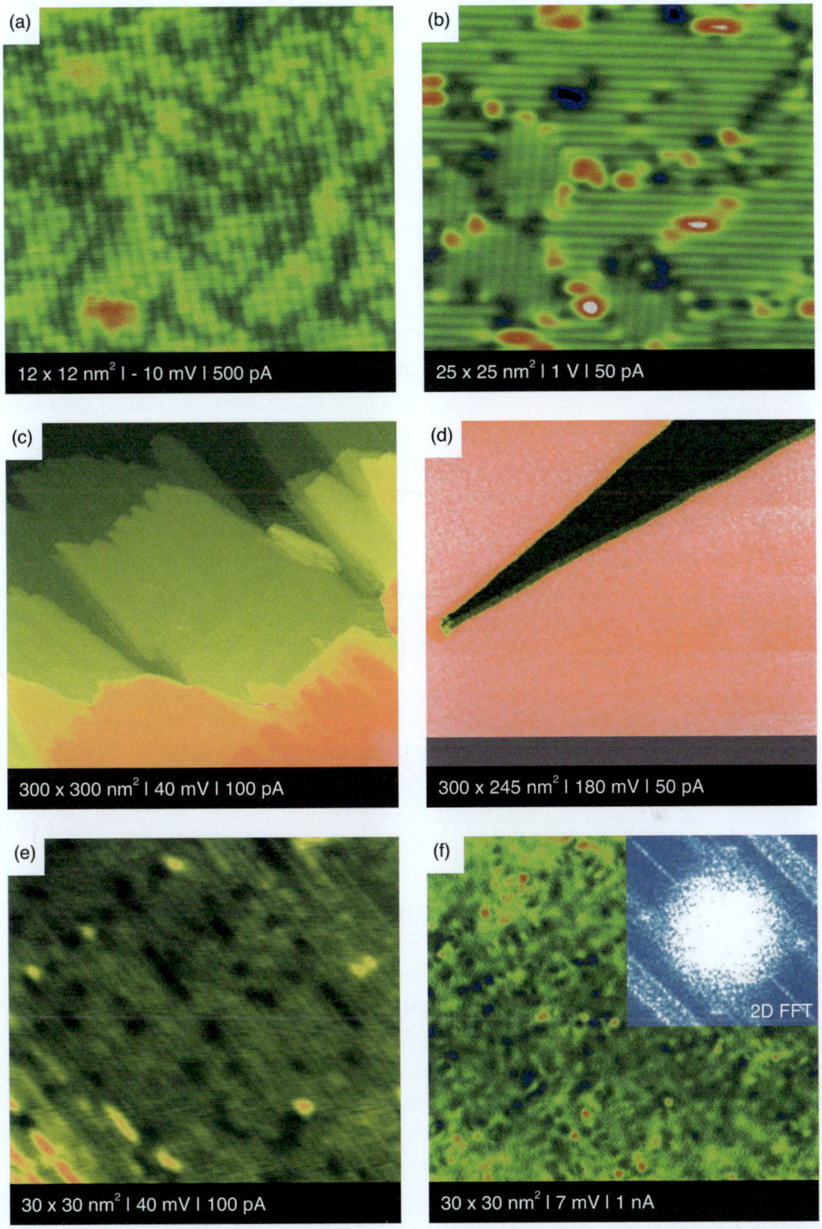

Figure 7.4: STM topography of different iron-based superconductors: (a) Atomic resolution on FeTe-Se50. (b) (2×1) superstructure on Ba122-Co6. (c) and (d): Large scale images of two different morphologies on the same sample of Sr122-P35. (e) and (f): closeups of the two different morphologies in (c) and (d), respectively.

the same experiment, though the second one was observed more often and was exclusively observed in other measurement sessions. In this regard attempts of explaining different surface morphologies only by the preparation parameters appear questionable. Future experiments, including alternative cleavage at room temperature, could elucidate the prerequisites for the different surface morphologies.

Tunneling spectroscopy

Scanning tunneling spectroscopy was performed on Ba122-Co6 as well as on Sr122-P35. In both cases a superconducting gap was observed as shown in Fig. 7.5(a) and (b), respectively. Like most other iron pnictides and in agreement with previous studies on Ba122-Co6 [75, 187] the differential conductance is U-shaped which is a characteristic of a fully gapped superconductor. That the conductance indeed vanishes around zero voltage was not observed in the previous works because they were performed at higher experimental temperatures. The magnitude of the gap, as roughly estimated from the position of the quasiparticle peaks, varies with position and is between 6 and 9 meV. This finding is in general agreement with a previous STM study which revealed position-dependent values of Δ between 4.5 and 8 meV with an average of $6.25 \pm 0.73\,\mathrm{meV}$ [75].

Further, the spectra in Fig. 7.5(a) show a position-depend substructure inside the gap. It could be indicative of the double gap behavior observed in an ARPES experiment on the same compound [188], which found two different gap values of 5.0 meV and 6.5 meV on the electron and hole pocket, respectively. This feature could not be resolved in the STM experiment by Yin *et al.* which was performed at 6.25 K [75] and also not in another STM study at 4.2 K [189]. In order to determine the value for the second gap and to identify its spatial distribution, however, the data of the measurement presented in Fig. 7.5(a) is not yet sufficient.

In contrast to the situation at Ba122-Co6, STS on Sr122-P35 reveals a pronounced V-shaped gap, indicative of nodes of the order parameter on the Fermi surface as explained in Sec. 7.1. Complementary experiments also support this explanation [179]. Thus Sr122-P behaves like its direct "brother" in the 122 family, Ba122-P, where nodal superconductivity was already demonstrated [179]. Also this spectrum exhibits a substructure inside the gap which shows up more clearly in its numerical derivative presented in the inset of Fig. 7.5(b). Again it could be a hint at multiple gaps in this material. The fact that the differential conductance does not reach zero is due to the finite measurement temperature of 0.83 K in combination with the V-shape of the gap. The evolution of the gap with temperature is illustrated in Fig. 7.5(c). This measurement shows how the V-shape gets blunt with increasing thermal energy. Furthermore, it roughly confirms the superconducting transition temperature of 33 K [179].

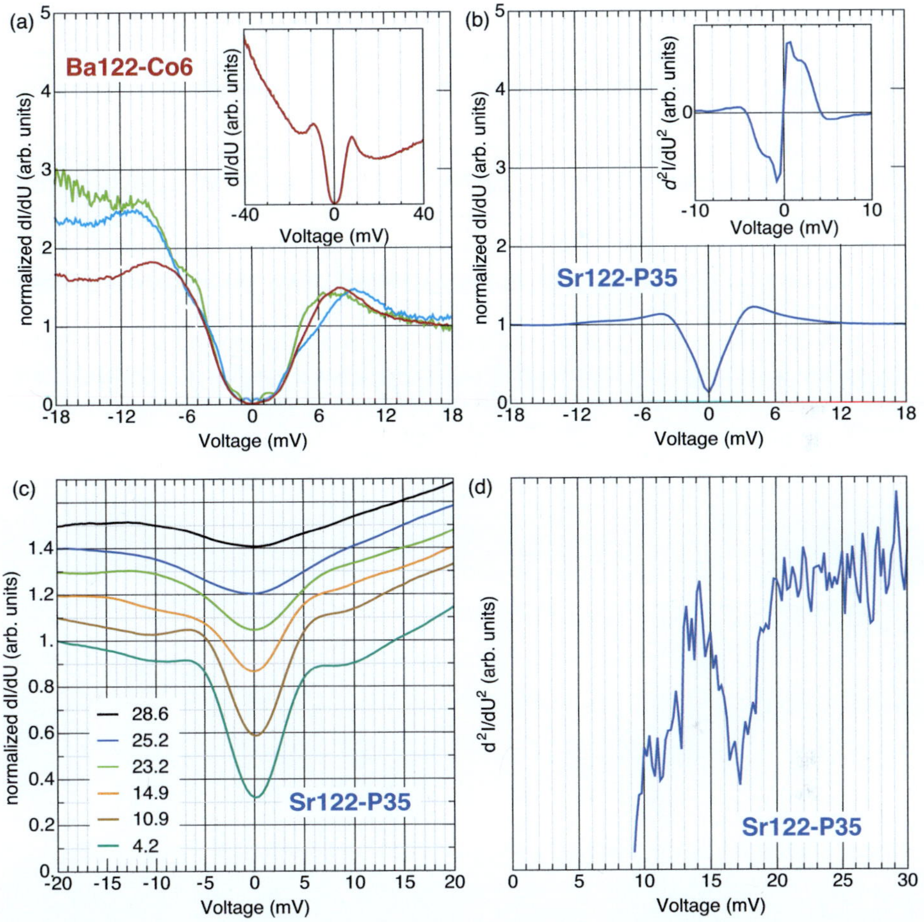

Figure 7.5: Tunneling spectroscopy on iron pnictide superconductors: (a) Spectra of fully gapped Ba122-Co6 (three different single positions, T=0.8 K). Gap magnitude, sub-gap structure and background conductance slightly vary with position. The inset shows the parabolic background conductance corresponding to the red curve on a larger voltage scale. (b) V-shaped spectrum of Sr122-P35 (average of many spectra, T=0.83 K). The inset shows the numerical derivative of this curve revealing a sub-gap structure. (c) Temperature dependent spectroscopy on Sr122-P35 (single position). (d) A dip in the 2f spectrum on Sr122-P35 may be suggestive of a bosonic mode.

Very promising with respect to the study of electron-boson coupling is the appearance of a wiggle at the negative side of the spectrum presented in Fig. 7.5(b). A second derivative spectrum recored later reveals a clear dip between 15 20 mV also on the positive bias side. The feasibility of observing this feature is facilitated by the lack of a large scale parabolic background as it is observed very often in unconventional superconductors (e. g. also shown in the inset of Fig. 7.5(a)). A conclusive statement whether this feature originates from a bosonic mode in Sr122-P35, however, needs more statistics to exclude tip effects and further assistance from complementary experiments like inelastic neutron scattering.

Despite the suboptimal surface morphology, the spectra on different loci did not vary much. In order to increase the quality of the data, however, the spectrum on Sr122-P35 was obtained by averaging about 2.5×10^5 spectra recorded in a grid of $60 \times 60\,\mathrm{nm}^2$. The same kind of grid measurement would, in principle, be the starting point for an analysis of the data in terms of QPI. Due to the present surface quality no spots could be detected in the 2D Fourier transforms except the ones arising from the lattice itself.

8 Conclusion and Outlook

The aim of the present work was the detailed study of electron-boson interactions in superconductors using tunneling spectroscopy in the geometry of the STM. At the beginning of this thesis it was pointed out that the crucial quantity for understanding the details of conventional phonon-mediated superconductivity is the Eliashberg function which is an effective phonon spectrum as it appears in the strong-coupling extension of BCS theory. The ongoing surprises concerning the discovery of novel superconducting compounds and the question for the "pairing glue" in these unconventional materials gave rise to an anew engagement in the aged and established concepts.

At the time when the Eliashberg theory was tested experimentally by electron tunneling spectroscopy the STM had not been available yet and thus the experiments were conducted on planar tunnel junctions involving a thin oxide layer instead of a true vacuum barrier. Nowadays the STM is commonly used in order to carry out tunneling spectroscopy, in particular also in the field of unconventional superconductivity where, in the recent years, a growing number of experimental observations was tried to be explained in terms of Eliashberg theory. A verification, that the old findings on planar tunnel junctions can be reproduced in the geometry of the STM, however, is not reported in literature.

Indeed, STM setups that provide the necessary energy resolution only started to be used more widely in the last few years. This is due to the fact that the technical requirements on a STM setup which operates at temperatures much below 4.2 K, the temperature of simply boiling helium, are demanding. The implementation in the design of the home-made JT-STM, which was used for the present experiments, was described in Chap. 4.

Furthermore, the signatures that can be expected from tunneling spectra with high energy resolution were explained, especially emphasizing the differences between tunneling in the superconducting state compared to tunneling in the normal state. While the first reflects the phonon characteristic in terms of an altered electronic density of states the latter directly yields the Eliashberg function in the second derivative of the tunneling current, though there exist controversial opinions on the last statement in literature. Therefore the results of a new analytical calculation concerning the inelastic contribution to the tunneling current were presented which confirm that

$$d^2I/dI^2 \propto \alpha^2F(\omega) \qquad \text{in the normal state,} \tag{8.1}$$

if the electronic densities of states are constant in the interval of interest and under some assumptions on the Fermi surfaces of both, the counter electrode and the sample. The latter might be softened in the geometry of the STM where momentum conservation during the tunneling process, as it is supposed in the calculation, does not have to be fulfilled.

The question to which extend inelastic processes contribute to the differential conductance when tunneling in the superconducting state still needs to be answered by theory. If it turns out that these effects cannot be neglected the previous studies which extracted the Eliashberg function from superconductor tunneling based on McMillan's inversion algorithm should be carefully reanalyzed.

The experiments on the conventional superconductors niobium and lead clearly revealed that phonon signatures can be indeed observed in tunneling spectra of the superconducting state in the geometry of the STM in full agreement with experiment on planar tunnel junctions. The main result of the present work, however, is that the phonons can also be addressed by low temperature STM-IETS when the superconductor is in its normal state allowing a more direct access to the Eliashberg function compared e. g. to McMillan's method. While similar conclusions had been drawn before (tough rarely) from experiments on planar junctions it is now possible to determine the Eliashberg function with real-space resolution.

In order to study phonons inelastically excited by the tunneling electrons in the normal state at very low temperatures superconductivity has to be suppressed. Two different approaches were presented - the application of a magnetic field and the inverse proximity effect, although the latter is restricted to superconductor films that are thinner than the respective coherence length.

As hypothesized by the prerequisite of a flat electronic density of states the Eliashberg function is much easier to obtain on a clean sample surface because adsorbates strongly influence the elastic (as well as the inelastic) conductance, especially in the low voltage range. Inelastic tunneling spectra on a niobium single crystal, which is known for its difficulties concerning the achievement of a pure surface, indeed showed the expected signatures at energies characteristic of van Hove singularities but only on lead islands, that could be prepared with much higher quality by MBE, the whole shape of the Eliashberg function could be resolved.

A general complication in STM-IETS experiments is the performance of the tip. Also there an adsorbate at the tip apex may have crucial influence on the resulting spectra. Unlike the situation in planar junctions where the tunneling current is averaged over the macroscopic contact area of both electrodes, in the case of the STM it only flows very locally and especially through the last atom or molecule on the tip. While averaging over the sample surface can be easily, tough time-consumingly, achieved by performing grid spectroscopy the influence of the tip has to be checked by changing the tip condition and repeating the measurement.

However, the STM is the only electron tunneling technique that allows to study unconventional superconductors since planar junctions cannot be fabricated in this case. Further, once a reliable tip condition is reached, the spatial resolution allows to study the Eliashberg function locally. In this way, a pronounced modification of the electron-phonon coupling strength was found in lead films of different thickness. In agreement with previous studies, that used different experimental approaches and focused also on the superconducting transition temperature, a bilayer oscillation was observed. The strongest enhancement takes place if an unoccupied electronic quantum well state lies closely above to the Fermi level. The reduced energetic stability of the respective layers is proposed to be responsible for the increased electron-phonon coupling. This interpretation is further verified by *ab initio* calculations for free-standing lead films. The enhancement mechanism, if generally applicable, could be used to increase the transition temperature of conventional superconductors by using layered structures with quantum well states at appropriate energies.

Different than initially expected and in contrast to the electronic system, no signs of phonons quantized in the direction perpendicular to the surface could be found even for the thinnest films. Therefore an extraction of the dispersion relation as it was performed by Gao *et al.* for magnons [53] was not possible and also no investigation of the momentum dependence of the electron phonon-coupling. In order to clarify whether the absence of phonon quantization has a physical reason or is due to a still insufficient energy resolution of the JT-STM preventing the observation of such quantization effects on the small energy scale given by the lead phonons a re-examination of this system at even lower temperatures, e. g. at the STM setup comprising a dilution refrigerator being built up at the moment, could by justified.

A prerequisite for successful measurements is the availability or attainability of suited samples. Due to this reason experiments on palladium-hydrogen (PdH) remained without results, most likely because, despite tremendous efforts, the hydrogen could not be trapped in the palladium lattice in a co-deposition preparation process at room temperature. Since superconductivity in this compound is caused by additionally introduced optical phonons [190] a study of the latter in the manner described above would be very interesting. Perhaps facilities that allow cooling of the substrate already during preparation which will also be available at the new dilution STM could help to overcome the difficulties in the preparation of suited PdH samples in the future.

An alternative to the *in situ* fabrication of samples is the use of externally grown crystals that are cleaved in UHV as it was carried out in this work in case of the experiments on iron-based superconductors. In order to accomplish the precedent preparation of the usually very small crystals a gadget was described, which increased the rate of successful cleaving attempts to 100 %. Nevertheless, there are almost no parameters which can be changed in the cleaving process itself and thus, concerning the resulting surface morphology one is at the mercy of the nature of the respective compound.

Preliminary results of scanning tunneling spectroscopy on two different "cousins" of the 122 family of iron pnictides, one of which (Sr122-P35) had not been investigated by STM before, seem to support the proposed $s^{\pm}$ symmetry of the order parameter which is able to explain the appearance of both, a gap function with nodes and one without nodes in the same class of compounds. Further, the very low temperature of the present experiment allowed to reveal a substructure inside the superconducting gaps which could be indicative for multiple gaps from different sheets of the Fermi surface. The next steps in this ongoing study should comprise the investigation of Sr122-P compounds of different doping level. For example spin-polarized STM could be able to reveal the magnetic order in the non-superconducting region of the phase diagram near the parent compound. Furthermore, the exceptional long standing time of the JT-STM should be exploited more, in order to obtain high-quality QPI maps, which unfortunately is only meaningful if a suited surface is obtained. Most relevant with respect to the focus of the present work, however, is the survey of the potential bosonic mode, that was observed in the superconducting state, when the sample is partly driven into the normal state by the application of a magnetic field.

On the whole, the experiments of the present work once more confirmed the accuracy of the concept of phonon-mediated superconductivity in the strong-coupling extension of BCS theory for conventional materials. There is also more and more evidence that spin fluctuations are the origin of Cooper pairing with a $s^{\pm}$ symmetry in the case of iron-based superconductors. Nonetheless, further intense research is justified since a conclusive theory is still missing. Following A. J. Leggett [191], the demand on such a theory, in the end, is to "provide [...] a blueprint for a robust room-temperature superconductor" or to state with confidence why room temperature superconductivity with all its imaginable fascinating applications has to remain a dream.

Bibliography

[1] W. Buckel and R. Kleiner, *Supraleitung: Grundlagen und Anwendungen* (Wiley-VCH, 2004).

[2] Siemens AG, "Magnetic resonance imaging," http://www.healthcare.siemens.de/magnetic-resonance-imaging (2014/04/06).

[3] CERN, "Pulling together: Superconducting electromagnets," http://home.web.cern.ch/about/engineering/pulling-together-superconducting-electromagnets (2014/04/06).

[4] T. Hime, P. A. Reichardt, B. L. T. Plourde, T. L. Robertson, C.-E. Wu, A. V. Ustinov, and J. Clarke, "Solid-state qubits with current-controlled coupling," Science, **314**, 1427 (2006).

[5] Leibniz Institute for Solid State and Materials Research Dresden (IFW Dresden), "Supratrans," http://www.supratrans.de/en/home/index.html (2014/04/06).

[6] J. G. Bednorz and K. A. Müller, "Possible high T_c superconductivity in the Ba-La-Cu-O system," Zeitschrift für Physik B Condensed Matter, **64**, 189 (1986).

[7] R. Gross and A. Marx, *Festkörperphysik* (Oldenburg, 2012).

[8] Y. Kamihara, T. Watanabe, M. Hirano, and H. Hosono, "Iron-based layered superconductor La[$O_{1-x}F_x$]FeAs ($x = 0.05 - 0.12$) with $T_C = 26$ K," Journal of the American Chemical Society, **130**, 3296 (2008).

[9] W. L. McMillan and J. M. Rowell, "Lead phonon spectrum calculated from superconducting density of states," Physical Review Letters, **14**, 108 (1965).

[10] H. Kamerlingh Onnes, "The superconductivity of mercury," Comm. Phys. Lab. Univ. Leiden, **122**, 124 (1911).

[11] W. Meissner and R. Ochsenfeld, "Ein neuer Effekt bei Eintritt der Supraleitfähigkeit," Naturwissenschaften, **21**, 787 (1933).

[12] F. London and H. London, "The electromagnetic equations of the supraconductor," Proceedings of the Royal Society of London A, **149**, 71 (1935).

[13] J. Bardeen, L. N. Cooper, and J. R. Schrieffer, "Theory of superconductivity," Physical Review, **108**, 1175 (1957).

[14] E. Maxwell, "Isotope effect in the superconductivity of mercury," Physical Review, **78**, 477 (1950).

[15] G. Czycholl, *Theoretische Festkörperphysik: von den klassischen Modellen zu modernen Forschungsthemen* (Springer DE, 2008).

[16] J. F. Annett, *Supraleitung, Suprafluidität und Kondensate* (Oldenbourg, München, 2011).

[17] J. R. Schrieffer, *What is a quasi-particle?* (National Bureau of Standards, 1970).

[18] N. W. Ashcroft and N. D. Mermin, *Festkörperphysik* (Oldenbourg, München, 2013).

[19] G. Grimvall, "The electron-phonon interaction in normal metals," Physica Scripta, **14**, 63 (1976).

[20] G. Eliashberg, "Interactions between electrons and lattice vibrations in a superconductor," Soviet Physics, **11**, 696 (1960).

[21] D. J. Scalapino, J. R. Schrieffer, and J. W. Wilkins, "Strong-coupling superconductivity. I," Physical Review, **148**, 263 (1966).

[22] J. R. Schrieffer, D. J. Scalapino, and J. W. Wilkins, "Effective tunneling density of states in superconductors," Physical Review Letters, **10**, 336 (1963).

[23] I. Giaever, H. R. Hart, and K. Megerle, "Tunneling into superconductors at temperatures below 1 °K," Physical Review, **126**, 941 (1962).

[24] W. L. McMillan, "Transition temperature of strong-coupled superconductors," Physical Review, **167**, 331 (1968).

[25] A. Schilling, M. Cantoni, J. Guo, and H. Ott, "Superconductivity above 130 K in the Hg-Ba-Ca-Cu-O system," Nature, **363**, 56 (1993).

[26] F. Steglich, J. Aarts, C. Bredl, W. Lieke, D. Meschede, W. Franz, and H. Schäfer, "Superconductivity in the presence of strong pauli paramagnetism: $CeCu_2Si_2$," Physical Review Letters, **43**, 1892 (1979).

[27] G. Wu, Y. Xie, H. Chen, M. Zhong, R. Liu, B. Shi, Q. Li, X. Wang, T. Wu, Y. Yan, *et al.*, "Superconductivity at 56 K in samarium-doped SrFeAsF," Journal of Physics: Condensed Matter, **21**, 142203 (2009).

[28] Z.-A. Ren, W. Lu, J. Yang, W. Yi, X.-L. Shen, Z.-C. Li, G.-C. Che, X.-L. Dong, L.-L. Sun, F. Zhou, *et al.*, "Superconductivity at 55 K in iron-based F-doped layered quaternary compound Sm[$O_{1-x}F_x$]FeAs," arXiv:0804.2053 (2008).

[29] H.-H. Wen and S. Li, "Materials and novel superconductivity in iron pnictide superconductors," Annual Review of Condensed Matter Physics, **2**, 121 (2011).

[30] D. Wu, N. Barišić, M. Dressel, G. H. Cao, Z.-A. Xu, E. Schachinger, and J. P. Carbotte, "Eliashberg analysis of optical spectra reveals a strong coupling of charge carriers to spin fluctuations in doped iron-pnictide $BaFe_2As_2$ superconductors," Physical Review B, **82**, 144519 (2010).

[31] L. Boeri, O. Dolgov, and A. Golubov, "Is $LaFeAsO_{1-x}F_x$ an electron-phonon superconductor?" Physical Review Letters, **101**, 026403 (2008).

[32] J. E. Hoffman, "Spectroscopic scanning tunneling microscopy insights into Fe-based superconductors," Reports on Progress in Physics, **74**, 124513 (2011).

[33] D. C. Johnston, "The puzzle of high temperature superconductivity in layered iron pnictides and chalcogenides," Advances in Physics, **59**, 803 (2010).

[34] D. J. Scalapino, "A common thread: The pairing interaction for unconventional superconductors," Review of Modern Physics, **84**, 1383 (2012).

[35] J. E. Hoffman, K. McElroy, D.-H. Lee, K. M. Lang, H. Eisaki, S. Uchida, and J. C. Davis, "Imaging quasiparticle interference in $Bi_2Sr_2CaCu_2O_{8+\delta}$," Science, **297**, 1148 (2002).

[36] E. L. Wolf, *Principles of electron tunneling spectroscopy*, International series of monographs on physics (Oxford Univ. Pr., New York, 1989).

[37] F. Schwabl, *Quantenmechanik (QM I): Eine Einführung*, 7th ed. (Springer Berlin Heidelberg, 2007).

[38] I. Giaever, "Electron tunneling and superconductivity," Science, **183**, 1253 (1974).

[39] J. C. Fisher and I. Giaever, "Tunneling through thin insulating layers," Journal of Applied Physics, **32**, 172 (1961).

[40] I. Giaever, "Energy gap in superconductors measured by electron tunneling," Physical Review Letters, **5**, 147 (1960).

[41] J. Nicol, S. Shapiro, and P. H. Smith, "Direct measurement of the superconducting energy gap," Physical Review Letters, **5**, 461 (1960).

[42] J. Bardeen, "Tunnelling from a many-particle point of view," Physical Review Letters, **6**, 57 (1961).

[43] C. J. Chen, *Introduction to scanning tunneling microscopy*, Oxford series in optical and imaging sciences ; 4 (Oxford Univ. Press, New York, 1993).

[44] P. Albrecht, K. Ritter, and L. Ruppalt, "The bardeen transfer hamiltonian approach to tunneling and its application to STM/carbon nanotubes," http://nanohub.org/resources/382 (2013/12/15).

[45] J. Tersoff and D. Hamann, "Theory and application for the scanning tunneling microscope," Physical Review Letters, **50**, 1998 (1983).

[46] J. A. Stroscio, R. Feenstra, and A. Fein, "Electronic structure of the Si(111) 2×1 surface by scanning-tunneling microscopy," Physical Review Letters, **57**, 2579 (1986).

[47] R. Feenstra, J. A. Stroscio, and A. Fein, "Tunneling spectroscopy of the Si(111) 2×1 surface," Surface Science, **181**, 295 (1987).

[48] R. C. Jaklevic and J. Lambe, "Molecular vibration spectra by electron tunneling," Physical Review Letters, **17**, 1139 (1966).

[49] B. C. Stipe, M. A. Rezaei, and W. Ho, "Single-molecule vibrational spectroscopy and microscopy," Science, **280**, 1732 (1998).

[50] A. J. Heinrich, J. A. Gupta, C. P. Lutz, and D. M. Eigler, "Single-atom spin-flip spectroscopy," Science, **306**, 466 (2004).

[51] T. Balashov, T. Schuh, A. F. Takács, A. Ernst, S. Ostanin, J. Henk, I. Mertig, P. Bruno, T. Miyamachi, S. Suga, and W. Wulfhekel, "Magnetic anisotropy and magnetization dynamics of individual atoms and clusters of Fe and Co on Pt(111)," Physical Review Letters, **102**, 257203 (2009).

[52] T. Schuh, T. Miyamachi, S. Gerstl, M. Geilhufe, M. Hoffmann, S. Ostanin, W. Hergert, A. Ernst, and W. Wulfhekel, "Magnetic excitations of rare earth atoms and clusters on metallic surfaces," Nano Letters, **12**, 4805 (2012).

[53] C. L. Gao, A. Ernst, G. Fischer, W. Hergert, P. Bruno, W. Wulfhekel, and J. Kirschner, "Spin wave dispersion on the nanometer scale," Physical Review Letters, **101**, 167201 (2008).

[54] T. Balashov, A. F. Takács, M. Däne, A. Ernst, P. Bruno, and W. Wulfhekel, "Inelastic electron-magnon interaction and spin transfer torque," Physical Review B, **78**, 174404 (2008).

[55] L. Vitali, M. A. Schneider, K. Kern, L. Wirtz, and A. Rubio, "Phonon and plasmon excitation in inelastic electron tunneling spectroscopy of graphite," Physical Review B, **69**, 121414 (2004).

[56] P. K. Hansma, "Inelastic electron tunneling," Physics Reports, **30**, 145 (1977).

[57] J. Klein, A. Léger, M. Belin, D. Défourneau, and M. J. L. Sangster, "Inelastic-electron-tunneling spectroscopy of metal-insulator-metal junctions," Physical Review B, **7**, 2336 (1973).

[58] J. M. Rowell, P. W. Anderson, and D. E. Thomas, "Image of the phonon spectrum in the tunneling characteristic between superconductors," Physical Review Letters, **10**, 334 (1963).

[59] D. J. Scalapino, "The electron-phonon interaction and strong-coupling superconductors," in *Superconductivity*, Vol. 1, edited by R. D. Parks (Dekker, New York, 1969) pp. 449–560.

[60] C. Brun, I.-P. Hong, F. Patthey, I. Y. Sklyadneva, R. Heid, P. M. Echenique, K. P. Bohnen, E. V. Chulkov, and W.-D. Schneider, "Reduction of the superconducting gap of ultrathin Pb islands grown on Si(111)," Physical Review Letters, **102**, 207002 (2009).

[61] A. Leger and J. Klein, "Experimental evidence for superconductivity enhancement mechanism," Physics Letters A, **28**, 751 (1969).

[62] J. M. Rowell, W. L. McMillan, and W. L. Feldmann, "Phonon emission and self-energy effects in normal-metal tunneling," Physical Review, **180**, 658 (1969).

[63] W. Wattamaniuk, H. Kruezer, and J. Adler, "Determination of phonon spectra of normal metals from inelastic electron tunneling," Physics Letters A, **37**, 7 (1971).

[64] J. G. Adler, H. J. Kreuzer, and W. J. Wattamaniuk, "Multichannel theory of inelastic electron tunneling in normal metal-insulator-metal junctions," Physical Review Letters, **27**, 185 (1971).

[65] M. Taylor, "Inelastic processes in tunnelling electrodes," Ultramicroscopy, **42–44, Part 1**, 215 (1992).

[66] P. Hlobil and J. Schmalian, "private communication," (2014).

[67] D. P. Smith, G. Binnig, and C. F. Quate, "Detection of phonons with a scanning tunneling microscope," Applied Physics Letters, **49**, 1641 (1986).

[68] Ø. Fischer, M. Kugler, I. Maggio-Aprile, C. Berthod, and C. Renner, "Scanning tunneling spectroscopy of high-temperature superconductors," Reviews of Modern Physics, **79**, 353 (2007).

[69] F. C. Niestemski, S. Kunwar, S. Zhou, S. Li, H. Ding, Z. Wang, P. Dai, and V. Madhavan, "A distinct bosonic mode in an electron-doped high-transition-temperature superconductor," Nature, **450**, 1058 (2007).

[70] C.-L. Song and J. E. Hoffman, "Pairing insights in iron-based superconductors from scanning tunneling microscopy," Current Opinion in Solid State and Materials Science, **17**, 39 (2013).

[71] L. Shan, J. Gong, Y.-L. Wang, B. Shen, X. Hou, C. Ren, C. Li, H. Yang, H.-H. Wen, S. Li, and P. Dai, "Evidence of a spin resonance mode in the iron-based superconductor $Ba_{0.6}K_{0.4}Fe_2As_2$ from scanning tunneling spectroscopy," Physical Review Letters, **108**, 227002 (2012).

[72] Z. Wang, H. Yang, D. Fang, B. Shen, Q.-H. Wang, L. Shan, C. Zhang, P. Dai, and H.-H. Wen, "Close relationship between superconductivity and the bosonic mode in $Ba_{0.6}K_{0.4}Fe_2As_2$ and $Na(Fe_{0.975}Co_{0.025})As$," Nature Physics, **9**, 42 (2013).

[73] W. Malaeb, T. Shimojima, Y. Ishida, K. Okazaki, Y. Ota, K. Ohgushi, K. Kihou, T. Saito, C. H. Lee, S. Ishida, M. Nakajima, S. Uchida, H. Fukazawa, Y. Kohori, A. Iyo, H. Eisaki, C.-T. Chen, S. Watanabe, H. Ikeda, and S. Shin, "Abrupt change in the energy gap of superconducting $Ba_{1-x}K_xFe_2As_2$ single crystals with hole doping," Physical Review B, **86**, 165117 (2012).

[74] T. Shimojima, W. Malaeb, K. Ohgushi, A. Chainani, S. Shin, S. Ishida, M. Nakajima, S. Uchida, T. Saito, H. Fukazawa, Y. Kohori, K. Kihou, C. H. Lee, A. Iyo, H. Eisaki, and K. Ishizaka, *Anomalous two peak structure in the Angle-resolved Photoemission Spectra of $Ba_{1-x}K_xFe_2As_2$*, arXiv e-print 1206.3163 (ISSP, University of Tokyo, 2012).

[75] Y. Yin, M. Zech, T. L. Williams, X. F. Wang, G. Wu, X. H. Chen, and J. E. Hoffman, "Scanning tunneling spectroscopy and vortex imaging in the iron pnictide superconductor $BaFe_{1.8}Co_{0.2}As_2$," Physical Review Letters, **102**, 097002 (2009).

[76] C. Ren, Z.-S. Wang, H.-Q. Luo, H. Yang, L. Shan, and H.-H. Wen, "Evidence for two energy gaps in superconducting $Ba_{0.6}K_{0.4}Fe_2As_2$ single

crystals and the breakdown of the Uemura plot," Physical Review Letters, **101**, 257006 (2008).

[77] L. Zhang, T. Miyamachi, T. Tomanić, R. Dehm, and W. Wulfhekel, "A compact sub-kelvin ultrahigh vacuum scanning tunneling microscope with high energy resolution and high stability," Review of Scientific Instruments, **82**, 103702 (2011).

[78] L. Zhang, *Sub-kelvin scanning tunneling microscopy on magnetic molecules*, Ph.D. thesis, Karlsruhe Institute of Technology (KIT) (2012).

[79] H. Ibach, *Physics of surfaces and interfaces* (Springer, 2006).

[80] K. Jousten and M. Wutz, eds., *Handbuch Vakuumtechnik: Theorie und Praxis* (Vieweg, Wiesbaden, 2006).

[81] F. Pobell, *Matter and methods at low temperatures*, 3rd ed. (Springer, Berlin, 2007).

[82] M. N. Wilson, *Superconducting magnets* (Clarendon Press, Oxford, 2002).

[83] Y. Iwasa, *Case studies in superconducting magnets : design and operational issues*, 2nd ed. (Springer, New York, 2009).

[84] Supercon, Inc, "NbTi superconducting wires," www.supercon-wire.com/content/nbti-superconducting-wires (2014/04/15).

[85] Supercon, Inc, "Certificate of compliance," (2013).

[86] J. Kirschner, H. Engelhard, and D. Hartung, "An evaporation source for ion beam assisted deposition in ultrahigh vacuum," Review of Scientific Instruments, **73**, 3853 (2002).

[87] J. Bostock and M. MacVicar, "Comment on the state-of-the-art of tunneling into superconducting niobium," Physics Letters A, **71**, 373 (1979).

[88] R. French, "Intrinsic type-2 superconductivity in pure niobium," Cryogenics, **8**, 301 (1968).

[89] WebElements: the periodic table on the WWW, "Niobium," http://www.webelements.com/niobium/ (2014/03/11).

[90] J. Bostock, V. Diadiuk, W. Cheung, K. H. Lo, R. Rose, and M. MacVicar, "Does strong-coupling theory describe superconducting Nb," Physical Review Letters, **36**, 603 (1976).

[91] Y. Nakagawa and A. D. B. Woods, "Lattice dynamics of niobium," Physical Review Letters, **11**, 271 (1963).

[92] L. Y. L. Shen, "Superconductivity of tantalum, niobium and lanthanum studied by electron tunneling: Problems of surface contamination," in *AIP Conference Proceedings*, Vol. 4 (AIP Publishing, 1972) pp. 31–44.

[93] E. Wolf, J. Zasadzinski, J. Osmun, and G. B. Arnold, "Proximity electron tunneling spectroscopy I. experiments on Nb," Journal of Low Temperature Physics, **40**, 19 (1980).

[94] J. D. Shore, M. Huang, A. T. Dorsey, and J. P. Sethna, "Density of states in a vortex core and the zero-bias tunneling peak," Physical Review Letters, **62**, 3089 (1989).

[95] T. Tomanic, *Untersuchung des elektronischen Oberflächenzustands von Ag-Inseln auf supraleitendem Niob (110)* (KIT Scientific Publishing, 2012).

[96] C. Sürgers, M. Schöck, and H. von Löhneysen, "Oxygen-induced surface structure of Nb(110)," Surface Science, **471**, 209 (2001).

[97] B. An, S. Fukuyama, K. Yokogawa, and M. Yoshimura, "Surface structures of clean and oxidized Nb(100) by LEED, AES, and STM," Physical Review B, **68**, 115423 (2003).

[98] I. Arfaoui, J. Cousty, and C. Guillot, "A model of the $NbO_{x\approx 1}$ nanocrystals tiling a Nb(110) surface annealed in UHV," Surface Science, **557**, 119 (2004).

[99] R. Franchy, T. Bartke, and P. Gassmann, "The interaction of oxygen with Nb(110) at 300, 80 and 20 K," Surface Science, **366**, 60 (1996).

[100] M. V. Kuznetsov, A. Razinkin, and A. L. Ivanovskii, "Oxide nanostructures on a Nb surface and related systems: experiments and ab initio calculations," Physics-Uspekhi, **53**, 995 (2010).

[101] M. Grundner and J. Halbritter, "On the natural Nb_2O_5 growth on Nb at room temperature," Surface Science, **136**, 144 (1984).

[102] A. Darlinski and J. Halbritter, "On angle resolved x-ray photoelectron spectroscopy of oxides, serrations, and protusions at interfaces," Journal of Vacuum Science & Technology A, **5**, 1235 (1987).

[103] J. Halbritter, "On the oxidation and on the superconductivity of niobium," Applied Physics A, **43**, 1 (1987).

[104] C. C. Koch, J. O. Scarbrough, and D. M. Kroeger, "Effects of interstitial oxygen on the superconductivity of niobium," Physical Review B, **9**, 888 (1974).

[105] A. Gurevich, "Enhancement of the upper critical field by nonmagnetic impurities in dirty two-gap superconductors," Physical Review B, **67**, 184515 (2003).

[106] H. Hess, R. Robinson, R. Dynes, J. Valles Jr, and J. Waszczak, "Scanning-tunneling-microscope observation of the Abrikosov flux lattice and the density of states near and inside a fluxoid," Physical Review Letters, **62**, 214 (1989).

[107] C. Renner, A. Kent, P. Niedermann, Ø. Fischer, and F. Lévy, "Scanning tunneling spectroscopy of a vortex core from the clean to the dirty limit," Physical Review Letters, **67**, 1650 (1991).

[108] C. P. Poole, ed., *Handbook of superconductivity* (Academic Press, San Diego, 2000).

[109] R. Heid, K.-P. Bohnen, I. Y. Sklyadneva, and E. V. Chulkov, "Effect of spin-orbit coupling on the electron-phonon interaction of the superconductors Pb and Tl," Physical Review B, **81**, 174527 (2010).

[110] M. M. Özer, J. R. Thompson, and H. H. Weitering, "Hard superconductivity of a soft metal in the quantum regime," Nature Physics, **2**, 173 (2006).

[111] T. Nishio, T. An, A. Nomura, K. Miyachi, T. Eguchi, H. Sakata, S. Lin, N. Hayashi, N. Nakai, M. Machida, and Y. Hasegawa, "Superconducting Pb island nanostructures studied by scanning tunneling microscopy and spectroscopy," Physical Review Letters, **101**, 167001 (2008).

[112] T. Cren, D. Fokin, F. Debontridder, V. Dubost, and D. Roditchev, "Ultimate vortex confinement studied by scanning tunneling spectroscopy," Physical Review Letters, **102**, 127005 (2009).

[113] Y. X. Ning, C. L. Song, Y. L. Wang, X. Chen, J. F. Jia, Q. K. Xue, and X. C. Ma, "Vortex properties of two-dimensional superconducting Pb films," Journal of Physics: Condensed Matter, **22**, 065701 (2010).

[114] D. Eom, S. Qin, M.-Y. Chou, and C. K. Shih, "Persistent superconductivity in ultrathin Pb films: A scanning tunneling spectroscopy study," Physical Review Letters, **96**, 027005 (2006).

[115] T. Nishio, M. Ono, T. Eguchi, H. Sakata, and Y. Hasegawa, "Superconductivity of nanometer-size Pb islands studied by low-temperature scanning tunneling microscopy," Applied Physics Letters, **88**, 113115 (2006).

[116] J. Noffsinger and M. L. Cohen, "First-principles calculation of the electron-phonon coupling in ultrathin Pb superconductors: Suppression of the transition temperature by surface phonons," Physical Review B, **81**, 214519 (2010).

[117] P. C. Hohenberg, "Existence of long-range order in one and two dimensions," Physical Review, **158**, 383 (1967).

[118] N. D. Mermin and H. Wagner, "Absence of ferromagnetism or antiferromagnetism in one- or two-dimensional isotropic Heisenberg models," Physical Review Letters, **17**, 1133 (1966).

[119] T. Zhang, P. Cheng, W.-J. Li, Y.-J. Sun, G. Wang, X.-G. Zhu, K. He, L. Wang, X. Ma, X. Chen, Y. Wang, Y. Liu, H.-Q. Lin, J.-F. Jia, and Q.-K. Xue, "Superconductivity in one-atomic-layer metal films grown on Si(111)," Nature Physics, **6**, 104 (2010).

[120] Y. Guo, Y.-F. Zhang, X.-Y. Bao, T.-Z. Han, Z. Tang, L.-X. Zhang, W.-G. Zhu, E. G. Wang, Q. Niu, Z. Q. Qiu, J.-F. Jia, Z.-X. Zhao, and Q.-K. Xue, "Superconductivity modulated by quantum size effects," Science, **306**, 1915 (2004).

[121] P. F. Newman and D. F. Holcomb, "Metal-insulator transition in Si:As," Physical Review B, **28**, 638 (1983).

[122] P.-G. de Gennes, *Superconductivity of metals and alloys* (Westview, Perseus Books, 1966).

[123] P.-G. de Gennes, "Boundary effects in superconductors," Reviews of Modern Physics, **36**, 225 (1964).

[124] P. Hilsch, "Zum Verhalten von Supraleitern im Kontakt mit Normalleitern," Zeitschrift für Physik, **167**, 511 (1962).

[125] R. Otero, A. L. Vázquez de Parga, and R. Miranda, "Can electron confinement barriers be determined by STM?" Surface Science, **447**, 143 (2000).

[126] Y.-F. Zhang, J.-F. Jia, T.-Z. Han, Z. Tang, Q.-T. Shen, Y. Guo, Z. Q. Qiu, and Q.-K. Xue, "Band structure and oscillatory electron-phonon coupling of Pb thin films determined by atomic-layer-resolved quantum-well states," Physical Review Letters, **95**, 096802 (2005).

[127] M. Jalochowski, H. Knoppe, G. Lilienkamp, and E. Bauer, "Photoemission from ultrathin metallic films: Quantum size effect, electron scattering, and film structure," Physical Review B, **46**, 4693 (1992).

[128] J. R. Anderson and A. V. Gold, "Fermi surface, pseudopotential coefficients, and spin-orbit coupling in lead," Physical Review, **139**, A1459 (1965).

[129] E. Ogando, N. Zabala, E. V. Chulkov, and M. J. Puska, "Self-consistent study of electron confinement to metallic thin films on solid surfaces," Physical Review B, **71**, 205401 (2005).

[130] J. Paggel, T. Miller, and T.-C. Chiang, "Quantum-well states as Fabry-Pérot modes in a thin-film electron interferometer," Science, **283**, 1709 (1999).

[131] J. H. Dil, J. W. Kim, S. Gokhale, M. Tallarida, and K. Horn, "Self-organization of Pb thin films on Cu(111) induced by quantum size effects," Physical Review B, **70**, 045405 (2004).

[132] A. Crottini, D. Cvetko, L. Floreano, R. Gotter, A. Morgante, and F. Tommasini, "Step height oscillations during layer-by-layer growth of Pb on Ge(001)," Physical Review Letters, **79**, 1527 (1997).

[133] W. B. Su, S. H. Chang, W. B. Jian, C. S. Chang, L. J. Chen, and T. T. Tsong, "Correlation between quantized electronic states and oscillatory thickness relaxations of 2D Pb islands on Si(111)-(7×7) surfaces," Physical Review Letters, **86**, 5116 (2001).

[134] M. C. Yang, C. L. Lin, W. B. Su, S. P. Lin, S. M. Lu, H. Y. Lin, C. S. Chang, W. K. Hsu, and T. T. Tsong, "Phase contribution of image potential on empty quantum well states in Pb islands on the Cu(111) surface," Physical Review Letters, **102**, 196102 (2009).

[135] K. Budde, E. Abram, V. Yeh, and M. Tringides, "Uniform, self-organized, seven-step height Pb/Si(111)-(7×7) islands at low temperatures," Physical Review B, **61**, R10602 (2000).

[136] R. Otero, A. L. Vázquez de Parga, and R. Miranda, "Observation of preferred heights in Pb nanoislands: A quantum size effect," Physical Review B, **66**, 115401 (2002).

[137] F. Schulte, "A theory of thin metal films: electron density, potentials and work function," Surface Science, **55**, 427 (1976).

[138] H. H. Weitering, D. R. Heslinga, and T. Hibma, "Structure and growth of epitaxial Pb on Si(111)," Physical Review B, **45**, 5991 (1992).

[139] A. Ishizaka and Y. Shiraki, "Low temperature surface cleaning of silicon and its application to silicon MBE," Journal of the Electrochemical Society, **133**, 666 (1986).

[140] S. H. Chang, W. B. Su, W. B. Jian, C. S. Chang, L. J. Chen, and T. T. Tsong, "Electronic growth of Pb islands on Si(111) at low temperature," Physical Review B, **65**, 245401 (2002).

[141] R. Negishi and Y. Shigeta, "Local structure and electronic state of a nanoscale Si island on Si(111)-7 × 7 substrate," Surface Science, **507–510**, 582 (2002).

[142] WebElements: the periodic table on the WWW, "Silicon," http://www.webelements.com/silicon/ (2014/04/27).

[143] K. Takayanagi, Y. Tanishiro, S. Takahashi, and M. Takahashi, "Structure analysis of Si(111)-7×7 reconstructed surface by transmission electron diffraction," Surface Science, **164**, 367 (1985).

[144] P. Auer and W. Mönch, "Cleaved si (111) surfaces: Geometrical and annealing behaviour," Surface Science, **80**, 45 (1979).

[145] J. Lander, G. Gobeli, and J. Morrison, "Structural properties of cleaved silicon and germanium surfaces," Journal of Applied Physics, **34**, 2298 (1963).

[146] S. Kitamura, T. Sato, and M. Iwatsuki, "Observation of surface reconstruction on silicon above 800°C using the STM," Nature, **351**, 215 (1991).

[147] WebElements: the periodic table on the WWW, "Copper," http://www.webelements.com/copper/ (2014/04/27).

[148] W. G. Moffatt, *The Handbook of Binary Phase Diagrams* (General Electric Comp., Schenectady, N.Y., 1978).

[149] C. F. Hirjibehedin, "Spin coupling in engineered atomic structures," Science, **312**, 1021 (2006).

[150] F. Leibsle, S. Dhesi, S. Barrett, and A. Robinson, "STM observations of Cu(100)-c(2×2)N surfaces: evidence for attractive interactions and an incommensurate c(2×2) structure," Surface Science, **317**, 309 (1994).

[151] T. Törndahl, *Atomic Layer Deposition of Copper, Copper(I) Oxide and Copper(I) Nitride on Oxide Substrates*, Ph.D. thesis, Uppsala Universitet (2004).

[152] F. M. Leibsle, C. F. J. Flipse, and A. W. Robinson, "Structure of the Cu(100)-c(2×2)N surface: A scanning-tunneling-microscopy study," Physical Review B, **47**, 15865 (1993).

[153] WebElements: the periodic table on the WWW, "Lead," http://www.webelements.com/lead/ (2014/04/27).

[154] I. B. Altfeder, K. A. Matveev, and D. M. Chen, "Electron fringes on a quantum wedge," Physical Review Letters, **78**, 2815 (1997).

[155] J. Camarero, J. Ferrón, V. Cros, L. Gómez, A. L. Vázquez de Parga, J. M. Gallego, J. E. Prieto, J. J. de Miguel, and R. Miranda, "Atomistic mechanism of surfactant-assisted epitaxial growth," Physical Review Letters, **81**, 850 (1998).

[156] C. Nagl, O. Haller, E. Platzgummer, M. Schmid, and P. Varga, "Submonolayer growth of Pb on Cu(111): surface alloying and de-alloying," Surface Science, **321**, 237 (1994).

[157] G. Kellogg and R. Plass, "The relationship between the growth shape of three-dimensional Pb islands on Cu(100) and the domain orientation of the underlying $c(\sqrt{5} \times \sqrt{5})r45$ structure," Surface Science, **465**, L777 (2000).

[158] C. Nagl, E. Platzgummer, O. Haller, M. Schmid, and P. Varga, "Surface alloying and superstructures of Pb on Cu(100)," Surface Science, **331-333**, 831 (1995).

[159] S. Robert, S. Gauthier, F. Bocquet, S. Rousset, J. L. Duvault, and J. Klein, "An STM study of the adsorption of Pb on Cu(100): formation of an ordered surface alloy," Surface Science, **350**, 136 (1996).

[160] F. Bocquet, S. Robert, S. Gauthier, J. Duvault, and J. Klein, "On the low-temperature growth of Pb on Cu(100)," Surface Science, **392**, 86 (1997).

[161] R. Plass and G. L. Kellogg, "Surface morphology changes during Pb deposition on Cu(100): evidence for surface alloyed Cu(100)-c(2×2)Pb," Surface Science, **470**, 106 (2000).

[162] X. Ma, P. Jiang, Y. Qi, J. Jia, Y. Yang, W. Duan, W.-X. Li, X. Bao, S. Zhang, and Q.-K. Xue, "Experimental observation of quantum oscillation of surface chemical reactivities," Proceedings of the National Academy of Sciences, **104**, 9204 (2007).

[163] Y. Ning, C. Song, Z. Guan, X. Ma, X. Chen, J. Jia, and Q. Xue, "Observation of surface superconductivity and direct vortex imaging of a Pb thin island with a scanning tunneling microscope," Europhysics Letters, **85**, 27004 (2009).

[164] M. M. Özer, J. R. Thompson, and H. H. Weitering, "Robust superconductivity in quantum-confined Pb: equilibrium and irreversible superconductive properties," Physical Review B, **74**, 235427 (2006).

[165] D. Saint-James and P.-G. de Gennes, "Onset of superconductivity in decreasing fields," Physics Letters, **7**, 306 (1963).

[166] K. Wang, X. Zhang, M. M. T. Loy, T.-C. Chiang, and X. Xiao, "Pseudogap mediated by quantum-size effects in lead islands," Physical Review Letters, **102**, 076801 (2009).

[167] A. Ernst, "private communication," (2014).

[168] G. Kresse and J. Furthmüller, "Efficient iterative schemes for ab initio total-energy calculations using a plane-wave basis set," Physical Review B, **54**, 11169 (1996).

[169] M. Lüders, A. Ernst, W. M Temmerman, Z. Szotek, and P. J. Durham, "Ab initio angle-resolved photoemission in multiple-scattering formulation," Journal of Physics: Condensed Matter, **13**, 8587 (2001).

[170] I. Y. Sklyadneva, R. Heid, K.-P. Bohnen, P. M. Echenique, and E. V. Chulkov, "Mass enhancement parameter in free-standing ultrathin Pb(111) films: The effect of spin-orbit coupling," Physical Review B, **87**, 085440 (2013).

[171] K. McElroy, R. W. Simmonds, J. E. Hoffman, D.-H. Lee, J. Orenstein, H. Eisaki, S. Uchida, and J. C. Davis, "Relating atomic-scale electronic phenomena to wave-like quasiparticle states in superconducting $Bi_2Sr_2CaCu_2O_{8+\delta}$," Nature, **422**, 592 (2003).

[172] M. L. Teague, G. K. Drayna, G. P. Lockhart, P. Cheng, B. Shen, H.-H. Wen, and N.-C. Yeh, "Measurement of a sign-changing two-gap superconducting phase in electron-doped $Ba(Fe_{1-x}Co_x)_2As_2$ single crystals using scanning tunneling spectroscopy," Physical Review Letters, **106**, 087004 (2011).

[173] T. Hänke, S. Sykora, R. Schlegel, D. Baumann, L. Harnagea, S. Wurmehl, M. Daghofer, B. Büchner, J. van den Brink, and C. Hess, "Probing the unconventional superconducting state of lifeas by quasiparticle interference," Physical Review Letters, **108**, 127001 (2012).

[174] M. Allan, A. Rost, A. Mackenzie, Y. Xie, J. Davis, K. Kihou, C. Lee, A. Iyo, H. Eisaki, and T.-M. Chuang, "Anisotropic energy gaps of iron-based superconductivity from intraband quasiparticle interference in LiFeAs," Science, **336**, 563 (2012).

[175] T. Hanaguri, S. Niitaka, K. Kuroki, and H. Takagi, "Unconventional s-wave superconductivity in Fe(Se,Te)," Science, **328**, 474 (2010).

[176] N. Katayama, S. Ji, D. Louca, S. Lee, M. Fujita, T. J. Sato, J. Wen, Z. Xu, G. Gu, G. Xu, *et al.*, "Investigation of the spin-glass regime between the antiferromagnetic and superconducting phases in $Fe_{1+y}Se_xTe_{1-x}$," Journal of the Physical Society of Japan, **79** (2010).

[177] M. Rotter, M. Tegel, and D. Johrendt, "Superconductivity at 38 K in the iron arsenide $(Ba_{1-x}K_x)Fe_2As_2$," Physical Review Letters, **101**, 107006 (2008).

[178] S. Nandi, M. G. Kim, A. Kreyssig, R. M. Fernandes, D. K. Pratt, A. Thaler, N. Ni, S. L. Bud'ko, P. C. Canfield, J. Schmalian, R. J. McQueeney, and A. I. Goldman, "Anomalous suppression of the orthorhombic lattice distortion in superconducting $Ba(Fe_{1-x}Co_x)_2As_2$ single crystals," Physical Review Letters, **104**, 057006 (2010).

[179] T. Kobayashi, S. Miyasaka, S. Tajima, T. Nakano, Y. Nozue, N. Chikumoto, H. Nakao, R. Kumai, and Y. Murakami, "Change of electronic state and crystal structure by postannealing in superconducting $SrFe_2(As_{0.65}P_{0.35})_2$," Physical Review B, **87**, 174520 (2013).

[180] T. Shimojima, F. Sakaguchi, K. Ishizaka, Y. Ishida, T. Kiss, M. Okawa, T. Togashi, C.-T. Chen, S. Watanabe, M. Arita, K. Shimada, H. Namatame, M. Taniguchi, K. Ohgushi, S. Kasahara, T. Terashima, T. Shibauchi, Y. Matsuda, A. Chainani, and S. Shin, "Orbital-independent superconducting gaps in iron pnictides," Science, **332**, 564 (2011).

[181] I. I. Mazin, D. J. Singh, M. D. Johannes, and M. H. Du, "Unconventional superconductivity with a sign reversal in the order parameter of $LaFeAsO_{1-x}F_x$," Physical Review Letters, **101**, 057003 (2008).

[182] C.-L. Song, Y.-L. Wang, P. Cheng, Y.-P. Jiang, W. Li, T. Zhang, Z. Li, K. He, L. Wang, J.-F. Jia, H.-H. Hung, C. Wu, X. Ma, X. Chen, and Q.-K. Xue, "Direct observation of nodes and twofold symmetry in FeSe superconductor," Science, **332**, 1410 (2011).

[183] W. Li, H. Ding, Z. Li, P. Deng, K. Chang, K. He, S. Ji, L. Wang, X. Ma, J.-P. Hu, *et al.*, "KFe_2Se_2 is the parent compound of K-doped iron selenide superconductors," Physical Review Letters, **109**, 057003 (2012).

[184] M. Naito and M. Hepp, "Superconducting T'-$La_{2-x}Ce_xCuO_4$ films grown by molecular beam epitaxy," Japanese Journal of Applied Physics, **39**, L485 (2000).

[185] K. Kihou, T. Saito, S. Ishida, M. Nakajima, Y. Tomioka, H. Fukazawa, Y. Kohori, T. Ito, S.-i. Uchida, A. Iyo, *et al.*, "Single crystal growth and characterization of the iron-based superconductor KFe_2As_2 synthesized by KAs flux method," Journal of the Physical Society of Japan, **79** (2010).

[186] F. Massee, S. de Jong, Y. Huang, J. Kaas, E. van Heumen, J. B. Goedkoop, and M. S. Golden, "Cleavage surfaces of the $BaFe_{2-x}Co_xAs_2$ and $Fe_ySe_{1-x}Te_x$ superconductors: A combined STM plus LEED study," Physical Review B, **80**, 140507 (2009).

[187] H. Zhang, J. Dai, Y. Zhang, D. Qu, H. Ji, G. Wu, X. F. Wang, X. H. Chen, B. Wang, C. Zeng, J. Yang, and J. G. Hou, "$\sqrt{2} \times \sqrt{2}$ structure and charge inhomogeneity at the surface of superconducting $BaFe_{2-x}Co_2As_2$ ($x = 0 - 0.32$)," Physical Review B, **81**, 104520 (2010).

[188] K. Terashima, Y. Sekiba, J. Bowen, K. Nakayama, T. Kawahara, T. Sato, P. Richard, Y.-M. Xu, L. Li, G. Cao, *et al.*, "Fermi surface nesting induced strong pairing in iron-based superconductors," Proceedings of the National Academy of Sciences, **106**, 7330 (2009).

[189] F. Massee, Y. Huang, R. Huisman, S. de Jong, J. B. Goedkoop, and M. S. Golden, "Nanoscale superconducting-gap variations and lack of phase separation in optimally doped $BaFe_{1.86}Co_{0.14}As_2$," Physical Review B, **79**, 220517 (2009).

[190] A. Eichler, H. Wühl, and B. Stritzker, "Tunneling experiments on superconducting palladium-deuterium alloys," Solid State Communications, **17**, 213 (1975).

[191] A. J. Leggett, "What DO we know about high T_c?" Nature Physics, **2**, 134 (2006).

Acknowledgments

I would like to express my gratitude to all the people without whom I would have never been able to arrive at this point.

I thank my doctorate supervisor **Prof. Wulf Wulfhekel** for giving me the opportunity for this doctoral research study and his trust in my ability to open up this new research field in his group. I especially appreciate his intense and direct guidance as well as the faithful care of all his students.

I thank **Prof. Jörg Schmalian** for being the second reviewer and for the interest that my work could arouse in him as a theoretician. Gratefully, I received his helpful comments on the theory of superconductivity.

I thank **Prof. Shigemasa Suga** for organizing a huge number of iron-based superconducting samples in his energetic attitude. Unfortunately, I could not yet examine all of them. Further, he showed me the way how top posts can be mounted in a controlled way.

I thank **Prof. Fumio Komori** for the possibility of visiting his group at the ISSP at the University of Tokyo. I am also grateful to all the people there, especially to Sunghun Kim, who assisted me in the experiments.

I thank the **KHYS** (Karlsruhe House of Young Scientists) for the financial support of my stay in Tokyo.

I thank **Dr. Toshio Miyamachi** for many conversations and discussions during many, many refreshment breaks, first in Karlsruhe and then in Tokyo, and for his friendship and his kind support during my time in Japan.

I thank **Dr. Christoph Sürgers**, also as representative for all colleagues at the PI, for getting us out of the jam in cases where the "Wulfhekels" were missing some essential parts or devices.

I thank the heads of the electronic and the mechanical workshop, **Roland Jehle** and **Michael Meyer**, and their teams, who never lost their nerves when I needed something to be done "now".

I thank **PD Arthur Ernst** and his co-workers for their supporting *ab initio* calculations and **Patrik Hlobil**, whose analytical calculations and explanations helped me a lot to gain deeper comprehension of the theory of inelastic tunneling.

I thank **Tobias Märkl**, **Florian Frech**, and **Jasmin Jandke**, the diploma and master students who successively joined me during the years, for their contribution. Special thanks goes to Jasmin and Tobias who, in the meantime, could be engaged for Ph.D. positions in our group, for proofreading the manuscript. In particular, I highly appreciate Tobias' thorough knowledge of English.

I thank **Dr. Timofey Balashov**, for his interest in basically everything you could ever be concerned with.

I thank **Moritz Peter** for his assistance in setting up the redesigned 4K-STM and also for his support in computer-related things as well as for sharing even the smallest rooms with me.

The atmosphere in the **AG Wulfhekel** deserves special appreciation. Going to work was a pleasure on almost every single day. Therefore my thank goes to all the students and especially to the long-time colleagues Stefan, Tobias, Lukas, Lei, Moritz, Tobias, Jasmin, Toyo, Timofey, Toshio, Marie, Hironari, and Wulf.

I am grateful to my **family** and especially to my parents for their support and their unshakable belief in me.

I thank **Harry** for reminding me, during the writing of the manuscript, of a walk outside which very often increased effectiveness afterwards.

Patiently waiting for me in Tirschenreuth, I thank **Caro** for her love.

Experimental Condensed Matter Physics
(ISSN 2191-9925)

Die Bände sind unter www.ksp.kit.edu als PDF frei verfügbar
oder als Druckausgabe bestellbar.

Band 1 Alexey Feofanov
Experiments on flux qubits with pi-shifters. 2011
ISBN 978-3-86644-644-1

Band 2 Stefan Schmaus
Spintronics with individual metal-organic molecules. 2011
ISBN 978-3-86644-649-6

Band 3 Marc Müller
Elektrischer Leitwert von magnetostriktiven Dy-Nanokontakten. 2011
ISBN 978-3-86644-726-4

Band 4 Torben Peichl
Einfluss mechanischer Deformation auf atomare Tunnelsysteme – untersucht mit Josephson Phasen-Qubits. 2012
ISBN 978-3-86644-837-7

Band 5 Dominik Stöffler
Herstellung dünner metallischer Brücken durch Elektromigration und Charakterisierung mit Rastersondentechniken. 2012
ISBN 978-3-86644-843-8

Band 6 Tihomir Tomanic
Untersuchung des elektronischen Oberflächenzustands von Ag-Inseln auf supraleitendem Niob (110). 2012
ISBN 978-3-86644-898-8

Band 7 Lukas Gerhard
Magnetoelectric coupling at metal surfaces. 2013
ISBN 978-3-7315-0063-6

Experimental Condensed Matter Physics
(ISSN 2191-9925)

Band 8 Kirill Fedorov
Fluxon readout for superconducting flux qubits. 2013
ISBN 978-3-7315-0067-4

Band 9 Jochen Zimmer
Cooper pair transport in arrays of Josephson junctions. 2014
ISBN 978-3-7315-0130-5

Band 10 Oliver Berg
Elektrischer Transport durch Nanokontakte von Selten-Erd-Metallen. 2014
ISBN 978-3-7315-0209-8

Band 11 Grigorij Jur'evic Grabovskij
Investigation of coherent microscopic defects inside the tunneling barrier of a Josephson junction. 2014
ISBN 978-3-7315-0210-4

Band 12 Cornelius Thiele
STM Characterization of Phenylene-Ethynylene Oligomers on Au(111) and their Integration into Carbon Nanotube Nanogaps. 2014
ISBN 978-3-7315-0235-7

Band 13 Michael Peter Schackert
Scanning Tunneling Spectroscopy on Electron-Boson Interactions in Superconductors. 2014
ISBN 978-3-7315-0238-8